More than two-thirds of stars belong to multiple stellar systems. Binary stars are considered now as one of the best constraints on stellar formation models. Not only do binaries keep memory of their birth conditions but their orbit will also be submitted to changes by tidal effects, wind accretion and encounters in clusters. Certainly the correlation between orbital eccentricity and period is a clue to our understanding of double star history. The most recent observations are discussed in these proceedings with the aim to disentangle evidence of stellar formation from later physical evolution. Each article is a paper that was read at a September 1991 meeting organized to honour Dr. Roger Griffin of the University of Cambridge for his pioneer work in galactic astronomy, dynamics of clusters and study on binary stars due to his cross-correlation technique to determine stellar radial velocities.

Binaries as Tracers of Stellar Formation

Frontispiece. The Aletsch glacier.

With a length of 18 km, this is one of the longest glaciers in the Alps. It is seen here from a ridge situated within a 30 mn walk from Bettmeralp, the small village where took place the meeting in September, 1991. One can note:
– the several central moraines (dark trails of stones on the glacier) originating from tributary glaciers;
– the limits of the last advance of the glacier (absence of vegetation on the mountain slopes near the glacier);
– the Jungfraujoch Observatory (marked by an arrow) on top of a small rocky peak, at 3600 m above sea-level.
– the heavy cloud coverage!

(From an assembly of pictures taken and kindly communicated by R. F. Griffin).

Binaries as Tracers of Stellar Formation

Edited by

Antoine Duquennoy
Observatoire de Genève

and

Michel Mayor
Observatoire de Genève

Published by the Press Syndicate of the University of Cambridge
The Pitt Building, Trumpington Street, Cambridge CB2 1RP
40 West 20th Street, New York, NY 10011–4211, USA
10 Stamford Road, Oakleigh, Victoria 3166, Australia

First published 1992

Printed in Great Britain at the University Press, Cambridge

A catalogue record for this book is available from the British Library

Library of Congress cataloguing in publication data available

ISBN 0 521 43358 4 hardback

Contents

Editors' note

In 1967, Roger Griffin published his first work (*ApJ 148, 465*) related to the determination of radial-velocities by cross-correlation technique. The exceptional efficiency of this method (a gain of more than 1000 times that of the photographic plate) allowed considerable progress during the past 25 years in many domains of stellar kinematics: statistical properties of binary stars, dynamics of globular clusters, galactic kinematics, etc.

In the domain of late-type double stars, progress is particularly remarkable. For the first time detailed orbital elements distributions are available for stars of different masses, ages and metallicities. It is now possible to search for traces of stellar formation among the statistical properties of binary stars. This workshop mainly showed (or recalled) the importance of the physical processes taking place *after* the stellar formation, which may alter the primordial properties of the binaries: tidal effects, angular momentum loss, mass accretion through stellar winds, dynamical interactions, etc.

Looking over the contributions presented here, the title of this workshop could have been " The $e - \log P$ Workshop". It also could have been "The 100th Paper Workshop". It is indeed to honour Roger Griffin that we organized this topical workshop, coinciding with Roger's 100th paper of his famous series published in *The Observatory*.

This meeting offered a real opportunity for intensive exchanges of ideas, during a week in the Swiss Alps (in Bettmeralp, a small village at an altitude of 1940 m above sea-level). The last paper of these proceedings, issued from the discussions, tries in particular to make a synthesis of the observations related to the orbital circularization of low-mass binary stars.

The Editors

A. Duquennoy   M. Mayor

Introductory Remarks

David W. Latham

The general concept of this workshop is to create an interaction between observation and theory; between the observers who are producing new information about the orbital characteristics of binaries and multiple stars in a variety of populations, and the theorists working on the formation and evolution of stars and stellar systems. The immediate excuse for holding this workshop in September 1991 is the expectation that Paper 100 in Roger Griffin's series "Spectroscopic Binary Orbits from Photoelectric Radial Velocities" will be published about two weeks from now. Just to make sure that I had my facts correct, I decided I had better go back and look through my personal collection of *The Observatory*. Paper I was published in the February 1975 issue, with B. Emerson as coauthor. The very first sentence reads:

> "The solution of the orbits of spectroscopic binaries has long been an important purpose of the measurement of stellar radial velocities."

Then, after a paragraph or two muttering about the importance of determining stellar masses, the paper outlines some of the history of the Cambridge effort:

> "Nearly all the observations have been made with the Cambridge 36-inch telescope in conjunction with the original radial-velocity spectrometer. Because of the heavy demand on that instrument for the purposes of narrow-band spectrophotometry for which it was originally designed, and the inconvenience of frequent alterations between the narrow-band and radial-velocity configurations, for several years the measurement of velocities was restricted to about three observing runs per annum. This was far from ideal for the observation of spectroscopic binaries having periods between a few months and a few years; and in any case the binary programme was not at that time regarded with high priority. Recently, however, changing patterns of telescope use have enabled the spectrometer to remain set up for comparatively long periods in its radial-velocity form, so that the observations have been much better distributed in time. Nevertheless, the twin constraints of telescope scheduling and local climate still conspire to prevent rapid progress in documenting any binary orbit, whatever its period. The situation may well, however, be regarded philosophically: the results of measuring almost any time-varying astronomical phenomenon are improved by increasing the time base, so this is one project where delay is of the essence!"

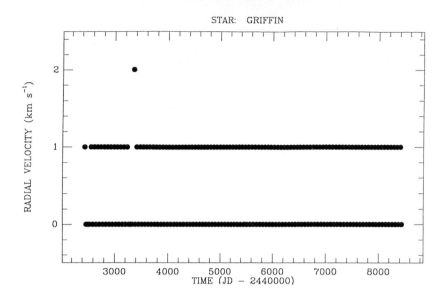

Figure 1. The time history of the publication of papers in the series "Spectroscopic Binary Orbits from Photoelectric Radial Velocities."

When I proceeded to look through the very next issue of *The Observatory*, for April 1975, I discovered that there was no Paper 2. For all these years I had been suffering the delusion that these papers had appeared one by one in every consecutive issue, like clockwork. I took some reassurance from finding Paper 2 in the next issue, June 1975, with Papers 3 through 13 following as expected. "Ah, it was just a slight hitch in getting started," I decided. But then, to my horror, I found another naked issue, June 1977. I took little comfort from the fact that both Papers 14 and 15 appeared in the following issue. With my confidence now badly shaken in the reliability of my expectation that Paper 100 would appear on 1 October 1991, I decided that I should do a proper analysis of the actual publication history.

In Figure 1 I plot the time history of publication. In a recent issue of *The Observatory* I found the statement that the nominal date of publication was the first of the month. So, to show that a paper was published, I assigned a velocity of 1 km s^{-1} to the corresponding Julian Day. *The Observatory* only appears every other month, so for the odd months in between I assigned a velocity of 0 km s^{-1}. You can see the gaps in April 1975 and June 1977 when no paper was published, and the anomalous issue of August 1977 when two papers appeared.

The power spectrum for these data, shown in Figure 2, has a very ugly window function, with quite a strong spurious peak near 20 days as well as the correct one near 60. Indeed, these data were rather cantankerous in resisting my efforts to find a satisfactory orbital solution, despite the fact that 98 points covering 198 months

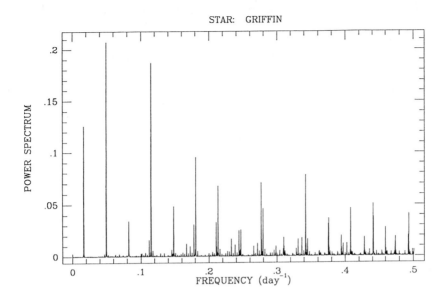

Figure 2. The power spectrum for the publication history shown in Figure 1.

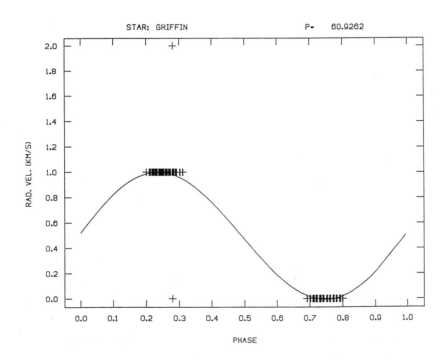

Figure 3. The orbital solution for the publication history.

were available (the August 1991 issue had not arrived at the other Cambridge by 3 September 1991). For example, Tsevi Mazeh's program found a curious beat solution with a period of almost exactly 60 days, when clearly this could not be correct, because the year is 365.25 days long, not 360. The best solution I could get without throwing away the two spurious points, shown in Figure 3, still gave a much better period, but a rather unsatisfactory error to the epoch.

To get serious again, you may rest assured that Paper 100 will indeed appear in the October 1991 issue.

But, this workshop is not just a Centennial Celebration for the long-running series of papers from Roger Griffin. It is really a celebration of the flowering of a branch of astronomical research that was pioneered by Roger and his collaborators; and I should acknowledge right away that among his collaborators not the least is another R. Griffin, who is also with us here in Bettmeralp - Cindy, it pleases me that you can join us in this mutual celebration. And, perhaps I should also mention Dr. G. A. Radford, who was a coauthor on Papers 3 through 17.

What is this flowering that I alluded to? I am not talking about blossoms that open in the morning after one night of observing. Instead we are now seeing the results of long-term research programs, some of them requiring as much as 10 or 20 years of observing before a proper analysis can be done. The point is that samples must be carefully selected, observed, and analyzed in order to take proper account of biases; in Roger Griffin's own words, from Paper 50, I give you the following example:

> "Therefore, despite the heavy bias still remaining against the discovery of binary systems having small amplitudes of velocity variation, a class of binaries which until now was grossly under-represented in the literature is gradually being documented ... It is no accident that the longest periods derived in the present series of papers are comparable with the length of the time which has elapsed since photoelectric observations began. By way of marking Paper 50, we put forward here the orbit with the longest period yet shown in this series of papers; but the periods of a number of systems still under observation are longer and still indeterminate."

and

> "For six years its [HD 185662's] velocity remained obstinantly constant - providing, incidently, an object lesson to those who might try too hastily to designate new radial-velocity standard stars. In 1976 there were at last slight signs of a change of velocity in the anticipated direction, and they were confirmed and amplified in 1977. In 1978, with majestic deliberation, the velocity passed the 1966 figure ..."

Or, from Paper 56:

"There are severe selection effects militating against the discovery of spectroscopic binaries with very high eccentricities. Such objects spend most of their time at velocities very close to the γ-velocity, and have short-lived and sometimes dramatic velocity excursions at periastron. Unless they are observed systematically over a long period of time, or else an observation happens to be made at a fortunate phase, the periastron passage can easily be missed and the small velocity changes at intervening times may be lost in the measurement noise. For large eccentricities the probability of discovery of a binary of given amplitude appears to be proportional to $(1 - e)$.

Even when a high-eccentricity binary has been discovered, there may be considerable difficulty in determining its orbit. Several revolutions can easily elapse before the observer is fully alerted to the abruptness of the periastron event or knows the period accurately enough to schedule intensive observations at the proper time; and when at last he *has* this information, it all too frequently happens that the next periastron passage occurs at the season when the object is lost in the daytime sky!"

So the sequence of 100 papers and Roger Griffin's dedication to the study of spectroscopic binaries is in several ways a model of the kind of approach that is needed for understanding the general characteristics of binaries, and how they formed, and how they can evolve.

Over the next two days we will hear about exciting new observational results on the characteristics of binaries in a wide variety of stellar populations: pre-main-sequence stars, main-sequence solar-type dwarfs, late-type main-sequence dwarfs, halo and thick-disk dwarfs, open cluster dwarfs, open cluster giants, field late-type giants, supergiants, and eclipsing binaries. Just about the only population that will not be covered is globular clusters.

And, I think you will find that the new observational results are bringing in to focus some tough challenges to the theoreticians. Can a general theory of star formation be developed that predicts the binary characteristics that we actually observe? What sets the initial distribution of secondary masses, orbital eccentricities, and periods? How do orbits evolve with time, both with and without mass transfer and mass loss? For example, can we understand orbital circularization well enough to use it as a clock for dating the age of coeval populations of binaries? Can we understand in detail the chemical abundances in systems where one or more of the stars has evolved and transferred mass? Can we model the dynamical evolution of clusters, and the role that binaries play in this evolution?

I do not expect answers to all these questions to emerge from this workshop, but as an observer I do hope that the theorists can give us some guidance about what are the key observations which must be made. And, in the meantime, we can continue to look forward to each new issue of *The Observatory*, and another paper in the series.

N-Body Simulations of Primordial Binaries and Tidal Capture in Open Clusters

Sverre J. Aarseth *

Abstract

We present results of N-body simulations pertaining to two aspects of open star cluster evolution. First we discuss the dynamics of a primordial population of circular binaries with short (10 to 100 days) periods. Although a large fraction of the binaries escape as a result of close encounters, the remaining population shows very little evidence of departure from zero eccentricity, suggesting that non-eccentric orbits above the observed cut-off period have a cosmogonic origin. The second part is concerned with computer modelling of tidal two-body capture. Provisional results indicate that dissipative close encounters may occur in high-density regions of open clusters containing a significant proportion of giants.

1. Introduction

Star clusters form ideal laboratories for testing the outcome of stellar evolution and mutual gravitational interactions. On the observational side there has been a big effort in establishing cluster membership. Radial velocities, which form the theme of the present meeting, play a vital part in this formidable task. On many occasions, stars which were thought to be single have been resolved using improved techniques to yield two close components with high orbital velocity (Griffin [Gri75]). By now, considerable evidence has accumulated that cluster binaries may be nearly as common as for the nearby field stars in both open clusters (Mathieu and Latham [ML86]; Mermilliod and Mayor [MM89]) and globular clusters (Pryor *et al.* [PCHF89]). Although observational selection favours the detection of shorter periods (< 1000 days), an open cluster such as $M67$ appears to have a binary frequency of 10% among its brightest members (Mathieu *et al.* [MLG90]). By analogy with the solar neighbourhood stars we may therefore expect a wide range in periods, corresponding to hard binaries which can survive the external perturbations of other stars in the cluster.

*Institute of Astronomy, University of Cambridge, Madingley Road, Cambridge CB3 0HA, England.

This 'molecular' nature of open clusters presents a considerable challenge for the dynamical experimentalist who is obliged to include interactions between the binaries and other cluster members in a more realistic treatment.

In the present paper, we concentrate on some aspects of the binary phenomenon in open clusters. Section 2 describes briefly the basis for the N-body simulation models. Taking the cue from recent determinations of the tidal circularization period in $M67$ (Latham *et al.* [LMMD92]), we attempt to shed light on the question of how long a population of relatively hard binaries may maintain an initial circular orbit in the presence of external perturbations. Section 3 also contains a general discussion of cluster evolution in the absence of mass loss from evolving stars. In Section 4 we consider the process of tidal two-body capture by close encounters. Although less important for systems composed of single stars, this type of binary formation may play a role in actual clusters containing hard binaries where the latter form compact subsystems by temporary capture and hence increase the probability of close encounters. Finally, Section 5 contains some discussions of the numerical results.

2. Simulation models of open clusters

Open star cluster dynamics is well defined and such systems are therefore quite amenable to numerical studies. The motion of most cluster stars are subject to a dominant gravitational force from the other members which holds the cluster together as it orbits the Galaxy. The perturbations due to the Galactic mass distribution exerts a smooth tidal force which is linear with respect to the cluster centre since typical cluster radii are very small compared to the size of the Galaxy. Assuming a circular orbit in the Galactic plane, we can then write the equations of motion for a mass-point m_i in rotating coordinates as

$$\frac{d^2 x_i}{dt^2} = F_x + 4A(A - B)\, x_i + 2\omega_z \frac{dy_i}{dt}, \frac{d^2 y_i}{dt^2} = F_y - 2\omega_z \frac{dx_i}{dt}, \frac{d^2 z_i}{dt^2} = F_z - C z_i. \quad (1)$$

Here the terms F_x, F_y, F_z denote the gravitational attraction due to the other $N - 1$ stars, A and B are Oort's Galactic constants, ω_z is the angular velocity and C is the force gradient in the vertical direction. These equations are also applicable to open clusters during small oscillations out of the Galactic plane.

The main computational effort of a cluster simulation is to obtain the sum of all individual attractions acting on a given star. A self-consistent treatment requires frequent updating of the total acceleration for each star over long time intervals in order to yield significant solutions. Even so, it is possible to follow the entire evolution of an open cluster with a few thousand members, using a direct solution method (see Aarseth [Aar85a] for full details of the numerical scheme). The main numerical problems of such simulations are connected with the need to study a variety of close encounters; *i.e.* between two single stars, a single star and a binary or two binaries. Such encounters are often very energetic and provide the main mechanism which drives the internal dynamical evolution. The smooth tidal field, on the other hand, exerts a more gentle influence; it lowers the escape barrier in the xy-direction, whereas the z-component gives rise to a tidal flattening. An irregular external perturbation, due to passing interstellar clouds, may also be included in the modelling; however, such effects appear to be quite small (Terlevich [Ter87]) and may be ignored here.

Initial cluster models with $N = 1000$ point-mass particles are generated from a Plummer distribution which has a modest central density concentration. The corresponding velocities are first selected from an equilibrium distribution (Section 3) and later (Section 4) taken to be nearly at rest, producing an early collapse which may be of interest for testing tidal two-body capture. For a realistic cluster simulation it is desirable to select individual stellar masses from a general mass function. Here we adopt a Salpeter-type power-law with exponent 2.35 and a modest mass ratio of 5:1 between the most massive and least massive star, and take the algebraic mean mass as $0.6\,m_\odot$. The effect of the tidal field is modelled by adopting a scaled length unit of 1 pc, corresponding to an equilibrium half-mass radius $R_h = 0.8$ pc. These cluster parameters, together with appropriate values of A, B and C then define an initial tidal radius $R_t = 11.9$ pc. Stars outside twice R_t are defined as having escaped from the cluster and are removed from the calculations; the resulting mass loss leads to a slow decrease of the tidal radius, determined from the first equation (1).

A population of initial binaries is now introduced as follows. Starting with the most massive member, every fifth star is subdivided into two equal-mass components of specified semi-major axis and eccentricity. This conservative choice of binary masses has been adopted in view of the lack of precise data for actual clusters. Hence the simulated binaries will not be subject to any preferential mass segregation compared to the massive single stars. We consider three models (denoted $M1$, $M2$, $M3$) with $N = 950$ initial single stars and a binary membership of $N_b = 50$, representing 5% of the total population or 20% of the 250 most massive stars. Bearing in mind that many of the observed binaries in $M67$ have rather shorter periods than the model binaries, this fraction may be a good compromise since the simulations do not include wide (but still hard) binaries. The models differ only in the choice of semi-major axis, with $a = 0.1, 0.22$ and $0.48\,A.U.$ in models $M1$, $M2$, $M3$. This corresponds to periods of about 10, 30 and 100 days, respectively, for the most massive binary of $1.6\,m_\odot$ and about 15, 45 and 150 days for the least massive binary. Since the main aim of these simulations is to place some limits on eccentricity changes due to external perturbations we adopt initial circular orbits for simplicity.

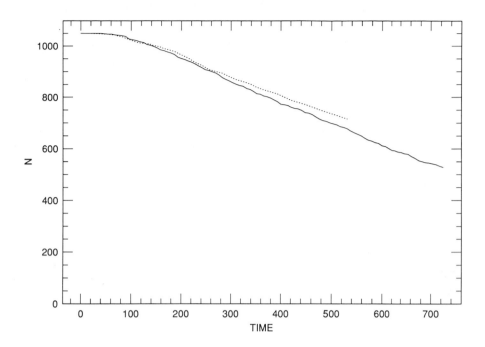

Figure 1. The number of cluster members as a function of time. Solid line: model $M2$; dotted line: model $M3$. Time is in units of 1×10^6 yrs.

3.   Dynamics of primordial binaries

The general aspects of cluster evolution in the presence of primordial binaries have been discussed in several recent papers (McMillian *et al.* [MHM90], [MHM91]; Heggie and Aarseth [HA92]), while other similar work has explored the observational consequences of runaway stars (Leonard and Duncan [LD88], [LD90]). In the present investigation we concentrate on the fate of the binaries, with special reference to the so-called period-eccentricity diagram which has featured prominently at this meeting. A key question here is concerned with the upper cut-off of the tidal circularization process and the extent to which this value may be affected by perturbations due to close encounters. The simulation models $M1$, $M2$, $M3$ do not include dissipative effects; hence any departures from circular binary orbits will be a direct measure of the gravitational perturbations and even small eccentricities will not be subject to the circularization process.

To illustrate the general behaviour of the numerical models, we show in Figure 1 the number of cluster members (single stars and binary components) inside $2R_t$ as a function of time, in units of 1×10^6 yrs, for models $M2$ and $M3$. Following a plateau during the early phase when the cluster expands to its tidal radius, the escape rate is nearly constant with a half-life $T_h \simeq 7 \times 10^8$ yrs (model $M2$), in reasonable accord with observed ages; *i.e.* *median* life-times of $\simeq 2 \times 10^8$ yrs. During this interval the population of hard binaries declines steadily from 50 to 35; this is somewhat slower than for the single stars because of their greater average mass.

Fixing our attention on the cluster binaries at the half-life epoch, 34 members retain their original components and the 35th forms a hierarchical triple with the inner binary also an original. Apart from the outer binary component (which has high eccentricity), only three of these binaries have an eccentricity which is non-zero to three significant figures, with respective values $e = 0.001, 0.008, 0.009$. Moreover, 28 out of the 34 primordial binaries retain their original sequential order which suggests they have not experienced any dominant perturbations; *i.e.* impact parameters comparable to the semi-major axis would normally involve switching to one of the special procedures. This simulation was continued until only $N = 148$ stars remained bound. At this stage (age 1.3×10^9 yrs) there were 17 binaries left, with 15 having zero eccentricity (to three figures) and two new rather soft binaries deviating significantly from circular motion.

An examination of the escaping binaries provides interesting clues about the type of interactions involved. At the half-life epoch, 17 hard binaries have been ejected from the cluster. Of these, 13 retain their original components with zero orbital eccentricity to three significant figures, whereas two have exchanged one companion each and one binary is a hierarchical triple with non-primordial components. However, these outcomes are not necessarily due to strong interactions. A total of 84 critical triple encounters and 26 encounters between two binaries took place during this interval. Such interactions are calculated by accurate regularization procedures which are only invoked if the impact parameter of an approaching particle is less than twice the binary semi-major axis (or the sum for two binaries) and there are no other significant perturbers. The nature of these interactions are confirmed by noting at least five well separated examples each of triple and quadruple encounters with minimum two-body separations $\simeq 3 \times 10^{-7}$, compared to an initial semi-major axis of 1.1×10^{-6} scaled length units (here 1 pc). This compares with minimum separations of $\simeq 2 \times 10^{-9}$ between single particles for models $M2$ and $M3$, corresponding to less than a tenth of the solar radius.

Closer inspection of the results show that 10 of the 13 primordial binaries which escape with zero eccentricity do not appear to have experienced any dominant encounters involving new regularizations. In these cases, the escape mechanism can be ascribed to more distant two-body (or multiple) encounters which may not affect the eccentricity. Thus the critical encounters referred to above are connected with temporary capture and/or exchange of components, as well as successive grazing encounters of hierarchical systems. Even so, non-dominant encounters involving very hard binaries may give rise to relatively high escape velocities.

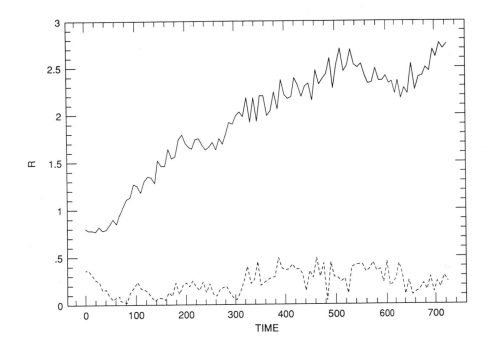

Figure 2. Half-mass radius and core radius as a function of time for model $M2$.

According to analytical theory (Heggie [Heg75]), the first-order perturbation cancels for circular orbits leaving a net effect of second order (as is also the case in general for the semi-major axis). This is due to the perturbing torque decreasing faster with increasing impact parameter for circular orbits than for eccentric orbits which have a permanent asymmetry (Hut and Paczynski [HP84]). The corresponding numerical cross sections confirm that the eccentricity change for near-circular orbits is significantly smaller than the analytical prediction for the general case. This weak dependence on the impact parameter may partly account for the relatively robust behaviour of the circular binaries studied here, with the choice of zero eccentricity motivated by astrophysical considerations.

It is also of interest to compare the spatial distribution of binaries in the simulations with actual observations. Figure 2 shows the time evolution of the half-mass radius and core radius in model $M2$. The former grows to about 2.5 pc which is similar to the observed (projected) value for single stars in $M67$ (Mathieu et al. [MLG90]), whereas the core radius remains below about 0.5 pc with considerable fluctuations. Although the combined masses of the simulated binaries are comparable to the other massive particles, they do exhibit a general mass segregation with respect to the total distribution. Thus the core defined by Figure 2 contains a fraction of massive binaries which is well in excess of the initial binary fraction of 5%, and to this extent

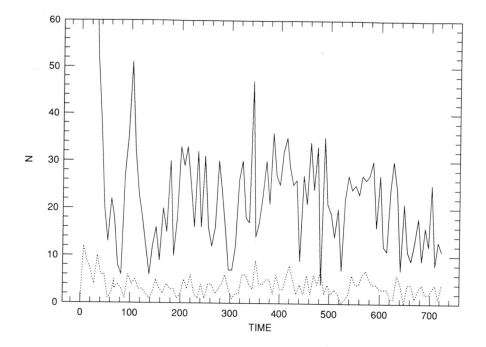

Figure 3. The total particle number and number of binaries inside one core radius as a function of time for model $M2$.

the binaries can be said to be centrally concentrated, in qualitative agreement with the observations [MLG90]. The corresponding number of particles (both types) and binaries in the core is displayed in Figure 3. During most of the evolution there are typically about 20 particles inside the core radius, of which four or five are binaries. Moreover, at half-life there are 13 circular binaries inside 1 pc. By this measure it appears that the simulation model shows a significant binary segregation.

For completeness we summarize some results of the binary behaviour in models $M1$ and $M3$. The first model was only studied for an interval of 1×10^8 yrs because of technical problems involving two hierarchical binaries. At this stage all the initial binaries remain with no noticeable eccentricity change, which is compatible with the results for the somewhat longer binary periods of model $M2$.

Model $M3$, which contains the widest binaries, also shows little evidence of eccentricity evolution. This model was terminated arbitrarily at an epoch of 5.3×10^8 yrs with a total cluster membership of $N = 716$ containing 32 hard binaries. Of the surviving binaries, 29 retain their original components with only one showing a significant eccentricity change ($e = 0.1$), while two distant binaries with exchanged components are in the process of escaping and the final binary has a semi-major axis of about $20\,A.U.$ with $e = 0.4$. This time 24 of the surviving binaries show no evidence of dominant perturbations (*i.e.* no regularization switching), and at the final epoch there are 9 circular binaries inside 1 pc. Moreover, among the 22 binaries which have escaped by this stage, 13 out of 14 retained their original components in circular orbits, and there were six energetic exchanges and two new but less hard binaries.

It may seem surprising that nearly all the remaining primordial binaries in these models do not show any significant departure from circular motion. However, even a period of $\simeq 100$ days corresponds to a relatively small value of the semi-major axis; *i.e.* $a = 0.48\,A.U.$ with the cluster size and mean mass adopted here. This should be compared with a typical two-body separation of $R_{cl} \simeq R_h/N \simeq 100\,A.U.$ used for treating close encounters by the Kustaanheimo-Stiefel regularization technique. Thus binaries escape more frequently as the result of encounters well outside the characteristic distance for significant internal changes ($\simeq 3\,a$), whereas most of the strong interactions lead to escape due to the recoil effect and sometimes in the case of two binaries, the destruction of one. Although the sling-shot type process can account for runaway stars from open clusters (cf. [LD90]), even non-dominant two-body encounters may produce escape velocities of binaries and single particles well in excess of the rms velocity.

4. Tidal capture

The presence of hard binaries in open clusters raises the possibility of very close approaches between the stars. One way in which this may occur is through a near-collision of two binaries, where there is a high probability of a hierarchical triple system (Mikkola [Mik83], [Mik84]) which may lead to a systematic increase of the inner binary eccentricity. Bound triples may also form when a single star is temporarily captured by a hard binary, as may occur during the post-collapse phase (Aarseth [Aar74], [Aar85b]; McMillan *et al.* [MHM91]). The relative importance of these processes depends on the number of binaries; hence the former case will be less important unless there is a substantial fraction of primordial binaries.

Since stars are not point-mass objects, approaches within a few stellar radii set up oscillations which may absorb sufficient energy for the orbital motion to become bound (Fabian *et al.* [FPR75]). Following more rigorous treatments of the two-body tidal capture process (Press and Teukolsky [PT77]; McMillian *et al.* [MDT87]; Ray *et al.* [RKA87]), we implement the resulting energy loss for the relative motion at pericentre, using semi-analytical fitting functions (Meinen and Portegies Zwart [MP91]). The tidal capture process is terminated when the binary semi-major axis reaches a prescribed value of about five stellar radii, to be associated with tidal synchronization. Two stars are merged into one object if their minimum impact parameter is less than 3/4 of their combined radii, as suggested by results of SPH simulations.

Table 1. Parameters for Tidal Two-Body Capture

$Model$	R_{ms}	N_{ms}	N_{ev}	R_{vir}	T_f	N_{sync}	N_{coll}
M9	10	1000	0	0.1	57	5	16
M10	1	1000	0	0.1	100	3	5
M11	50	200	800	0.5	58	7	8
M12	50	200	800	1.0	56	9	6
M13	100	200	800	0.5	34	6	6
M14	1	1000	0	1.0	93	2	2
M15	2	1000	0	1.0	327	1	1

In this exploratory investigation of tidal capture we mainly consider models without primordial binaries for simplicity. All models have a total membership of $N = 1000$ equal-mass particles, except for model $M15$ which has a Salpeter-type power-law mass function with mass ratio of 5 and mean mass of $1\,m_\odot$. The table summarizes the input parameters and main outcomes. Here Columns 1 and 2 denote the model name and corresponding stellar radius in units of $r_\odot = 0.005\,A.U.$, whereas Columns 3 and 4 show the number of main sequence and evolved stars, respectively, where the latter are defined to have zero radius. The equilibrium scale factor, R_{vir} (pc), is shown in Column 5, followed by the total time interval T_f in Column 6 expressed in units of the standard crossing time. In all these models the initial virial theorem ratio $T/V = 0.01$, which gives rise to a collapse ($T/V = 0.5$ for equilibrium).

The total number of synchronous binaries and collisions are given in Columns 7 and 8, respectively. Although the half-mass radius shrinks by a factor of $\simeq 3$ during the first few crossing times, there are no tidal capture events or collisions at this early stage. It is hardly surprising that there has been a significant number of close two-body interactions in models $M9$ and $M10$ which are characterized by small half-mass radii. Moreover, a number of multiple collisions occur; this can be understood in terms of increased central concentration of heavy bodies by mass segregation. Increasing the stellar radii as well as the half-mass radius in model $M11$ still gives rise to considerable dissipative interactions. Note that although only 200 of the stars have been assigned finite radii, their interactions with stars of zero radii inside about $150\,r_\odot$ can still result in tidal capture since the energy loss is merely halved. In fact, none of the eight collisions are between two stars of finite size, whereas four of these events involve a star which has already been merged. A similar number of events occur in model $M12$ which has a larger half-mass cluster radius, and now there are four collisions between the finite-size stars. Rather surprisingly, hyperbolic encounters leading to tidal capture do occur a few times in these models; however, some of these events are associated with temporary triple systems.

Increasing the stellar radii to $100 \, r_\odot$ in model $M13$ leads to an enhanced production of tidal binaries and merged objects during a relatively short interval, indicating a runaway accretion process in the core which would be suppressed by mass loss in a more realistic simulation. It should be noted that some of the tidal capture binaries counted in the table subsequently undergo collisions. Such an outcome is not the end product of the tidal capture process itself, since the pericentre of successive close interactions actually increases slowly because of angular momentum conservation. Instead these events usually occur after the formation of hierarchical triple systems where the outer pericentre is sufficiently small to perturb the inner binary eccentricity, which may have interesting observational consequences.

Models $M14$ and $M15$ are included to illustrate the behaviour of systems with relatively small stellar radii. In model $M14$, all the stars have one solar mass, with radii of $r_\odot$, and a population of $N_b = 50$ primordial binaries has been included in the manner of Section 2. Finally, model $M15$ contains a spectrum of single masses distributed in the range 2.7 to 0.5 $m_\odot$, and all having appropriate main-sequence radii (2 to 0.5 $r_\odot$). As expected, there is much less tidal two-body interaction in these models. Nevertheless, the results indicate that such processes may be relevant for many open clusters which have a wide range of masses where a significant proportion of the stars have evolved to the giant stage.

5.   Conclusions

The present study of primordial binaries was inspired by the apparent observational correlation between age and the maximum period for circular binaries due to the tidal circularization process (Latham *et al.* [LMMD92]). Using a fairly realistic N-body simulation of open star clusters, we have demonstrated that circular binaries with periods of up to $\simeq 100$ days do not increase their eccentricity significantly during the cluster lifetime. These results therefore support a cosmogonic interpretation for the observed transition to higher eccentricities.

Computer modelling of tidal two-body capture is still in its infancy. The present tentative results suggest that close two-body interactions are likely to occur at some stage during the evolution of open star clusters. In particular, we have found that hierarchical systems act as catalysts for promoting two-body dissipation and stellar collisions. However, the question of whether such processes will be sufficiently effective to have observable consequences can only be answered by more realistic simulations which are already under way. Thus the introduction of full stellar evolution in the numerical scheme is likely to reveal competing effects since mass loss from evolving stars will tend to prevent core collapse, counteracting the increased cross sections of giant-type stars.

References

[Aar74] Aarseth, S.J., 1974. *Astron. Astrophys. 35, 237*

[Aar85a] Aarseth, S.J., 1985a. In *Multiple Time Scales ed. Brackbill, J.U. and Cohen, B.I., Academic Press, p. 325*

[Aar85b] Aarseth, S.J., 1985b. In *Dynamics of Star Clusters, ed. Goodman, J. and Hut, P., D. Reidel, Dordrecht, p. 251*

[FPR75] Fabian, A.C., Pringle, J.E. and Rees, M.J., 1975. *MNRAS 172, 15P*

[Gri75] Griffin, R.F., 1975. *The Observatory 95, 23*

[HA92] Heggie, D.C. and Aarseth, S.J., 1992. *MNRAS (in press)*

[Heg75] Heggie, D.C., 1975. *MNRAS 173, 729*

[HP84] Hut, P. and Paczynski, B., 1984. *Astrophys. J., 284, 675*

[LMMD92] Latham, D.W., Mathieu, R.D., Milone, A.A.E. and Davis, R.J., 1992. *In Evolutionary Processes in Close Binaries, ed. R.S. Polidan, Kluwer, Dordrecht (in press)*

[LD88] Leonard, P.J.T. and Duncan, M.J., 1988. *Astron. J. 96, 222*

[LD90] Leonard, P.J.T. and Duncan, M.J., 1990. *Astron. J. 99, 608*

[ML86] Mathieu, R.D. and Latham, D.W., 1986. *Astron. J. 92, 1364*

[MLG90] Mathieu, R.D., Latham, D.W. and Griffin, R.F., 1990. *Astron. J. 100, 1859*

[MDT87] McMillan, S.L.W., Dermott, P.N. and Taam, R.E., 1987. *Astrophys. J. 318, 261*

[MHM90] McMillan, S., Hut, P. and Makino, J., 1990. *Astrophys. J. 362, 522*

[MHM91] McMillan, S., Hut, P. and Makino, J., 1991. *Astrophys. J. 372, 111*

[MP91] Meinen, T. and Portegies Zwart, S., 1991. Unpublished.

[MM89] Mermilliod, J.C. and Mayor, M., 1989. *Astron. Astrophys. 219, 125*

[Mik83] Mikkola, S., 1983. *MNRAS 203, 1107*

[Mik84] Mikkola, S., 1984. *MNRAS 207, 115*

[PT77] Press, W.H. and Teukolsky, S.A., 1977. *Astrophys. J. 213, 183*

[PCHF89] Pryor, C., McClure, R.D., Hesser, J.E. and Fletcher, J.M., 1989. *In Dynamics of Dense Stellar Systems* ed. D. Merritt, Cambridge University Press, p. *175*

[RKA87] Ray, A., Kembhavi, A.K. and Antia, H.M., 1987. *Astron. Astrophys. 184, 164*

[Ter87] Terlevich, E., 1987. *MNRAS 224, 193*

When and How can Binary Data test Stellar Models?

Johannes Andersen and Birgitta Nordström *

Abstract

The use of empirical data for components of binary systems (i.e. masses, radii, luminosities, composition) to test models for the evolution of stars from their formation to the present is briefly reviewed. The accuracy of the data, and its proper exploitation in the analysis, are of paramount importance. Suitable test techniques are discussed on the basis of recent examples.

1.   Introduction

With most active star forming regions hidden in dark clouds, our ideas of the properties of newborn stars are based on models of their early evolution as derived from stellar evolution theory. Because binary stars provide our only means (apart from the Sun) to empirically determine the most important input parameter to a stellar model - the *mass* - binary stars are key test objects for stellar evolution theory. Data of the required accuracy are currently only obtained from detached double-lined eclipsing binaries with favourable properties. Such systems also yield accurate empirical determinations of stellar *radius*, the most direct diagnostic of the evolution of a star of known mass.

Until we succeed in finding suitable binary systems to give good empirical masses and radii for true pre-main-sequence stars, our best targets are stars on or near the main sequence. From such systems, we can learn much about the processes going on in normal stars and the (in)adequacy of our models for their evolution. In the following, we shall briefly review the status of binary tests of stellar evolution models, with the caveat that substantial mass accretion during the contraction phase may make extrapolation of standard main-sequence models to earlier phases a dubious exercise,

---

*Copenhagen University Observatory, Brorfeldevej 23, DK-4340 Tølløse, Denmark

2. The data: what is needed, and what exists?

If one analyses the magnitude of the observable effects on parameters such as radius, colour, or luminosity which correspond to those model features of greatest current interest (opacities, convection treatment, metal and helium abundance), one finds that an accuracy of 1-2% in the mass and radius are required for a critical test. To match the potential of such data, calibrated colour indices on suitable photometric systems and high-quality spectroscopic metal abundance determinations are also needed.

Obtaining data of such quality is not a trivial enterprise. We shall not enter into the technical details here, since the subject has recently been thoroughly reviewed by Andersen [And91]. That paper also lists the resulting data for 45 eclipsing binary systems meeting the adopted acceptance criterion (mass and radius of both components known in a fundamental manner to ±2% or better). For a few of these systems, spectroscopic metallicity determinations are also available. Various applications of the data are discussed in more detail than is appropriate in the present context, and extensive references are given. Data used in the following discussion can be found in that review.

3. Techniques for testing stellar models

3.1. Historical

A test of theoretical stellar models with data from eclipsing binary systems was discussed already in the classic text by Schwarzschild [Sch58]. There, unevolved (ZAMS) models were compared in the mass-luminosity diagram with observed, early-type and presumably unevolved binary components. The agreement (within about two stellar magnitudes!) was considered gratifying at the time, but lack of actual information on evolutionary state and chemical composition left considerable ambiguity in the choice of stellar models.

Popper et al. [PJM70] plotted well-observed A and F-type binary components in the mass-radius diagram in order to, first, eliminate evolved stars and, second, use the locus of the unevolved stars to constrain the metal abundance of the models. Subsequently plotting the unevolved stars in the mass-luminosity diagram then allowed [PJM70] to derive the initial helium abundance of the stars. Still, there was no direct spectroscopic check of the derived heavy-element abundance (presumed uniform over the sample).

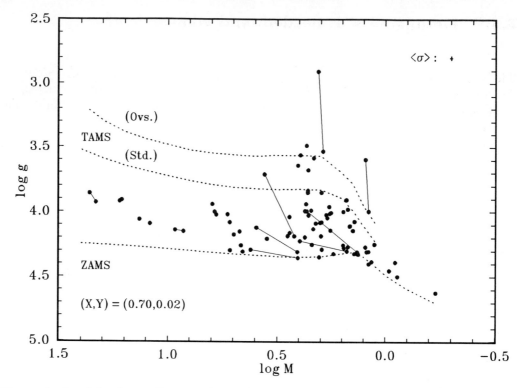

Figure 1. $\log M - \log g$ diagram for well-observed binary components. Data from [And91]; models from [ClG89], [ClG91]. Average error bars are shown

3.2.   Current status for stars in the main-sequence band.

In the above analyses, main-sequence stars were assumed to form a single-valued, sharply defined relation in diagrams displaying the various basic parameters. The progress since that time is dramatically displayed by the $\log M - \log g$ diagram of Fig.1, which shows the main-sequence as a wide band with a continuous distribution of stars, resolved to some ±5% of its width (note the size of the error bar!). The separation of stars of the same mass, but different ages (and perhaps composition) is clearly seen and require proper consideration in the comparison with models. Note that evolutionary tracks in this diagram are simple vertical lines (assuming no mass loss).

Already the basic data mass and radius (here used in the combination $\log M, \log g$) may provide a first test of the theoretical models: For two stars in the same binary system (a few examples are shown connected by lines), the models should simultaneously predict the correct radius for *both* stars for a single age. If the metal abundance of the system is known, a direct, non-trivial test of the selected models is possible; if not, at least the range of acceptable models can be narrowed down.

3.3.   Detailed fits to individual systems.

The most stringent test of the models is, of course, obtained when a minimum of free parameters are allowed in a fit to a maximum (=all!) of the observational constraints. The test outlined above does not confront the models with the full range of available information: The effective temperatures (or, equivalently, luminosities) of the stars have not been considered. The real litmus test is obtained when all of the parameters mass, radius, effective temperature, and (metal) abundance are fit to within observational errors of $\pm 1\%$ or less, at a single age for both stars in the system.

This blind test was first successfully performed for the F-type evolved system AI Phe by Andersen et al. [ACG88]. In the fit, the mixing-length parameter of the VandenBerg [VdB83], [VdB85] evolution models was fixed by matching a solar model to the observed Sun. Next, the helium abundance Y was determined from a fit to the luminosity of the less-evolved star, using the spectroscopic value of the metallicity parameter Z. The value found, $Y = 0.27 \pm 0.01$, was identical to that derived for the solar model. The age of the less-evolved star, and thus of the system, is derived simultaneously in the fit.

Subsequently, and thus with *no* adjustable parameters left, a model for the more massive star is calculated, using the precise observed mass, radius, temperature, and metal abundance, and with the exact same mixing length, helium abundance, and age used to fit the secondary. The rather spectacular fit obtained is shown in Fig.2. Subsequently, successful fits have been made for other systems spanning a metallicity range of a factor of two, from AI Phe up to the Hyades.

One of these systems, TZ For (Andersen et al. [ACN91]), turned out to have a number of especially informative properties. A satisfactory fit to the data was only obtained using models with a moderate amount of convective overshooting. Additional evidence in favour of this interpretation (but missed in the paper) is the fact that the more massive star would have exceeded its Roche lobe if it had suffered the helium flash prescribed by standard (but not overshooting) models for stars of this mass.

We shall not enter a lengthy discussion of the controversial subject of overshooting here, but do wish to mention below some of the observational facts that must be taken seriously in such discussions. The recent clarification that the former great increase in the age of $1.5 - 2 \mathcal{M}_\odot$ stars has turned out to be a numerical artifact of some early overshooting models (Schaller et al. [SSM92]) has not removed the need for a proper understanding of convection in stellar interiors.

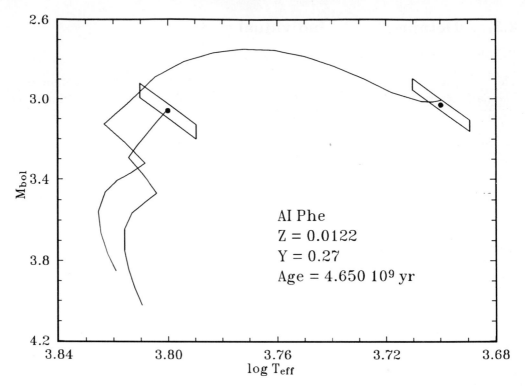

Figure 2. Model fit to the components of AI Phe. Data and error polygons from Andersen et al. [ACG88].

3.4.   Lifetime arguments.

One of the key diagnostics of convective core overshooting is the width of the main sequence. In an earlier paper (Andersen et al. [ANC90]), it was noted that the fraction of systems in the mass range $1.5 - 2.5\mathcal{M}_\odot$ which are placed by standard models in the rapid evolutionary phases on the blue edge of the Hertzsprung gap (beyond the "red hook" which marks central hydrogen exhaustion in stars with convective cores) far exceeded that expected from the respective lifetimes. As this excess could not be removed by plausible adjustments to the metal abundances (not all of which have yet been determined), the data were interpreted as strong evidence for the significance of convective overshooting in this mass range.

Stothers and Chin [StC91] plotted the evolved binary systems from our 1990 paper in the mass-radius and HR diagrams together with models without overshooting, but with new opacity data and a rather high metal abundance (Z = 0.03). They showed that most of the data points then fall between the ridge lines defined by ZAMS and TAMS models with little or no overshooting. From that, they concluded that the excess of evolved stars described above, and by implication the effect of overshooting, were likely not real.

Clearly, when new models with updated physics become available, conclusions obtained from earlier models may change. However, no actual evidence exists to prove or suggest that $Z = 0.03$ is appropriate to any or all of the observed systems. And because evolutionary tracks curve back into the main sequence band between the "red hook" phase and the Hertzsprung gap, a star may be in a rapid phase of evolution even when below the TAMS line in these diagrams: Only an actual fit to the individual stars with models for the precise observed masses can assign them to the correct segment of their evolutionary tracks. By omitting that step one ignores data which contain essential information.

3.5. Tidal interactions.

Tidal effects (apsidal motion, orbital circularization, rotational synchronization) offer another type of probe into stellar interiors. The time scales for these effects depend extremely sensitively on the stellar dimensions (radii, in particular). Hence, to usefully test theoretical predictions of orbital or rotational properties as a function of time, the physical dimensions of the stars must be known very accurately. Thus, we are led again to the sample of stars compiled by Andersen [And91], who also discussed their tidal properties. Without recapitulating that discussion in extenso, let us recall the illustrative case of TZ For. The system is unusual by having a circular orbit of long period (76^d) coupled with a non-synchronized secondary star, because synchronization is normally expected to proceed $\simeq 100$ times faster than circularization. The riddle is solved once it is realized that the red-giant primary star has passed the tip of the red giant branch rather recently. Because of its large radius at that stage, the star was able to circularize the binary orbit by itself very efficiently without affecting the rotation of its companion.

3.6. Isochrone fits.

A single binary system defines only two points on an isochrone. In contrast, from colour-magnitude diagrams of star clusters one may determine the shape of an isochrone in great detail (although the precise masses of the individual cluster stars are unknown). Binary stars play a role even here, albeit in a more negative sense: If the cluster diagram is severely contaminated by unrecognized binary (and non-member) stars, it may no longer be representative of the single cluster stars which the theoretical models describe, and isochrone fits may give misleading results. Some of these cases may be rather subtle: In the recent fit to the cluster diagram for NGC 3680 by Castellani et al. [CCS92], a handful of stars which are crucial to the fit with a standard model are shown by our extensive, unpublished radial-velocity data to be just binaries (and non-members). The true isochrone looks very different, as suggested by [ANC90].

4.   Directions for the future

The tests outlined above need to be expanded to other masses, chemical compositions, and ages than possible from the current data. Of most interest to the present meeting, more stars on (or preferably even before) the ZAMS need to be identified and studied. In addition, the chemical composition of more systems must be determined to remove the ambiguity of another free parameter from the tests. Finally, a few (open) star clusters in the right age and metallicity range need to be very carefully surveyed for membership and duplicity in order to complement the precise individual binary data with detailed isochrone fits. All these programmes are under way.

Acknowledgements. We thank the organizers for inviting us to such a pleasant and inspiring meeting, and Drs. J.V. Clausen and A. Giménez for a long and fruitful collaboration. Our work on eclipsing binaries is supported by observing time awarded by ESO and the Danish Board for Astronomical Research; additional financial support from the Danish Science Research Council and the Carlsberg Foundation is gratefully acknowledged.

References

[And91] Andersen, J. *1991, Astron. Astrophys. Review, 3, 91*

[ACG88] Andersen, J., Clausen, J.V., Gustafsson, B., Nordström, B., and Vanden-Berg, D.A. *1988, Astron. Astrophys. 196, 128*

[ACN91] Andersen, J., Clausen, J.V., Nordström, B., Tomkin, J., and Mayor, M. *1991, Astron. Astrophys. 246, 99*

[ANC90] Andersen, J., Nordström, B., and Clausen, J.V. *1990, Astrophys. J. 363, L33*

[CCS92] Castellani, V., Chieffi, A., and Straniero, O. *1992, Astrophys. J. Suppl. 78, 517*

[ClG89] Claret, A. and Giménez, A. *1989, Astron. Astrophys. Suppl. 81, 1*

[ClG91] Claret, A. and Giménez, A. *1991, Astron. Astrophys. Suppl. 87, 507*

[PJM70] Popper, D.M., Jørgensen, H.E., Morton, D.C., and Leckrone, D.S. *1970, Astrophys. J. 161, L57*

[SSM92] Schaller, G., Schaerer, D., Meynet, G., and Maeder, A. *1992, Astron. Astrophys. Suppl., in press*

[Sch58] Schwarzschild, M. *1958, Structure and Evolution of the Stars, Princeton Univ. Press, Princeton*

[StC91] Stothers, R.C. and Chin, C.-w. *1991, Astrophys. J. 381, L67*

[VdB83] VandenBerg, D.A. *1983, Astrophys. J. Suppl. 51, 29*

[VdB85] VandenBerg, D.A. *1985, Astrophys. J. Suppl. 58, 711*

Statistical Analysis of Single-Lined Red Giant Spectroscopic Binaries

Henri M.J. Boffin [*]   *Guy Paulus* [†]   *Nicolas Cerf* [‡]

Abstract

This article reports a statistical analysis of the orbital elements (period, eccentricity, mass function) performed on a sample of 194 single-lined red giant spectroscopic binaries. From the eccentricity-period diagram, we deduce a circularization cut-off period of 70 days and we find an eccentricity-period correlation that seemingly does not result from an observational bias. With the aid of two complementary methods, we derive the mass ratio distribution. A comparison with the observed mass function distribution shows a reasonably good agreement in the case of an uniform mass ratio distribution function and a single-valued giant mass of 1.5 $M_\odot$. We finally present a comparison with a sample of barium (Ba) stars.

1.   Introduction

The correct understanding of the binary properties exhibited by some classes of Peculiar Red Giants (PRG) like Ba, CH or S stars calls for a comparison with the characteristics of normal field giant binary systems. An analysis of the properties of a large sample of spectroscopic binaries (SB) containing at least one G or K giant is the subject of the present work. Although many statistical studies of SB have already been carried out, they deal with samples that either do not overlap or contain too few systems, or both. The sample of 194 SB1 systems to which we hereafter focus our attention is part of a more complete catalogue of late-type giant binaries (LTGB) that will be published in [BCP92].

---

[*]Institut d'Astronomie et d'Astrophysique, CP−165, Université Libre de Bruxelles, B−1050 Bruxelles, Belgium.

[†]Institut d'Astronomie et d'Astrophysique, CP−165, Université Libre de Bruxelles, B−1050 Bruxelles, Belgium, and Institut d'Astrophysique de Paris, 98bis bd Arago, F−75014 Paris, France. This author was partially supported by the Programme International de Coopération Scientifique (PICS N°18) and by the Science Programme of the EC (SC1 − 0065).

[‡]Institut d'Astronomie et d'Astrophysique, CP−165, Université Libre de Bruxelles, B−1050 Bruxelles, Belgium. This author is Research Assistant of the National Fund for Scientific Research (FNRS, Belgium).

2.   Distribution of orbital elements

Table 1. Mean properties of red giant binaries

	SB1	Ba
$\langle P \rangle$	1076	1413
$\langle \log P \rangle$	2.46	3.
$\langle e \rangle$	0.23[†]	0.14
$\langle e^2 \rangle$	0.1[‡]	0.04
$\langle f(m) \rangle$	0.13	0.025
$\langle \log f(m) \rangle$	−1.428	−1.92
[†] 0.3, [‡] 0.14 if $P \geq 100$ d		

In this Table, we give the mean values of the orbital elements for our sample of SB1 systems. Also shown for comparison are the mean values of Ba stars orbital elements ([McCW90]; [JM92]). Throughout this article, periods P are expressed in days and mass functions $f(m)$ in $M_\odot$.

2.1.   Period distribution

Because the period (P) varies over a wide range, it is more convenient and more rigorous to study the distribution of the logarithm of the period. The number of systems increases with $\log P$ until $\log P \gtrsim 3.5$, at which point a sharp jump occurs. This obviously results from a selection effect, as very large period systems are difficult to detect and a long time is needed to ascertain their orbits. This distribution is compatible with the one obtained by Duquennoy and Mayor [DM91] for field G-dwarfs, with very few short period systems ($\log P \lesssim 0.5$). For $0.5 \lesssim \log P \lesssim 3.5$, we may approximate the period distribution by the linear relation

$$f(\log P) \, \mathrm{d} \log P \; = \; 0.1666 \, \log P \, \mathrm{d} \log P \quad . \tag{1}$$

2.2.   Eccentricity distribution

The SB1 eccentricity distribution displayed in Fig. 1 clearly shows a substantial number of systems with nearly circular orbits. This is supposedly the result of tidal circularization. This effect can be removed by considering only systems with $P >$ 100 d, which is on the order of the cut-off period for this sample (see Sect. 2.3). The eccentricity distribution thus obtained is also shown in Fig. 1. The number of systems with $e < 0.1$ has considerably decreased, although still being non-zero. This may appear a puzzling point, and we will consider it further in Sect. 2.3. For $P > 100$ d, the deduced eccentricity distribution is not incompatible with

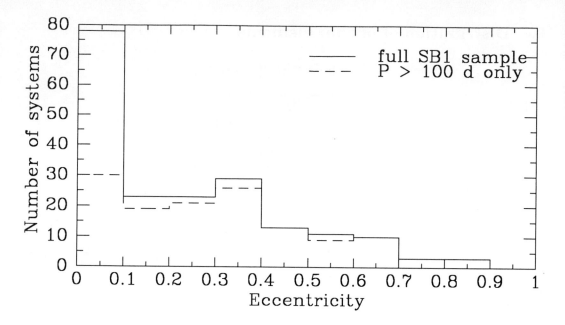

Figure 1. Eccentricity distributions for the whole SB1 sample (solid line) and for the sub-sample of SB1 with $P > 100$ d (dashed line).

$$f(e)\,\mathrm{d}e \;=\; 2\,(1 - e)\,\mathrm{d}e \quad , \tag{2}$$

which is coherent with what we could expect from the detection probability of a system (see e.g. [Tri90]).

2.3.   Eccentricity versus period diagram

2.3.1.   Circularization

The e versus $\log P$ diagram helps to better understand the process of tidal circularization. That theory (e.g. [Zah77]) indeed predicts that systems with orbital periods shorter than a certain cut-off period (which depends on the age of the system, as well as on the mass and radius of the giant) will be circularised. For example, Mermilliod and Mayor (this volume) derive, from their sample of open cluster red giant binaries, $\log P_{\text{cut-off}} \simeq 2.2$. Of course, in our sample the situation is not as clear, as we consider field giants and therefore have a priori a dispersion of ages and masses. More than one cut-off period might accordingly exist. At first glance, the $e - \log P$ diagram plotted in Fig. 2 seems to confirm that hypothesis.

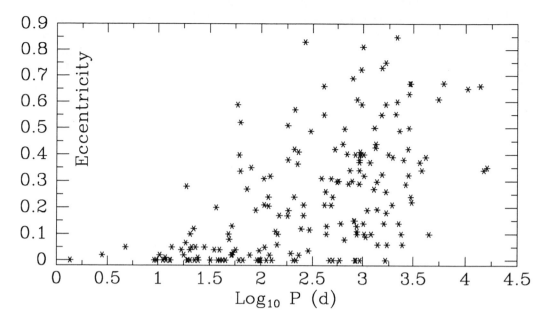

Figure 2. Eccentricity$-\log P$ diagram for all the stars contained in our catalogue [BCP92].

We nevertheless believe we may deduce with some confidence from Fig. 2 a single cut-off period at $\log P_{\text{cut-off}} \simeq 1.7$, i.e. a period three times smaller than the one reported by Mermilliod and Mayor (this volume). Below this cut-off period, all systems but three have an eccentricity below 0.1 . Those three stars undoubtedly appear to be giants; they might of course just have reached the giant branch and therefore have had no time to circularise. On the other hand, the spread in mass and age that might be present in our sample constitutes a plausible explanation for the existence of circular orbit systems with periods up to 150 days.

2.3.2.   Eccentricity-period correlation

The $e-\log P$ diagram displays another interesting feature. Indeed, we clearly see the scatter in eccentricity to be much greater for larger periods. This pattern presumably attests the existence of a correlation between e and P, which is found in various samples. To emphasize this relation we have fitted the mean value of the eccentricity, $\langle e \rangle$, as a function of $\log P$ for our sample. The relation turns out to be nicely linear, with

$$\langle e \rangle = 0.15 \, (\log P - 1) \quad .\tag{3}$$

We claim that this linear correlation does not result from an observational bias. To prove our assertion, we devise a very simple model. Assuming that a star is observed twice within a time lapse Δt, we compute the probability that the radial velocity V differs by more than ΔV. To this effect we write $V = KF$, with the amplitude K such that $K^3 = 2\pi G f(m) (1 - e^2)^{-3/2} P^{-1}$ and the variable part $F = (1 - e^2) \cos E (1 - e \cos E)^{-1}$. The eccentric anomaly E is related to time by $E - e \sin E = 2\pi t/P$. One can calculate the probability by numerically integrating

$$\Pi = \frac{1}{P} \int_0^P \Theta \left(\frac{K |F(t + \Delta t) - F(t)|}{\Delta V} - 1 \right) dt \quad , \tag{4}$$

where $\Theta(x)$ stands for the Heaviside step function. We emphasize at this point that our simple model is valid only when $\Delta t \lesssim P$. If it is not the case another method, described below, must be used. The computation has been made using $\Delta V \simeq 2 \, \mathrm{km \, s^{-1}}$ and $f(m)$ equal to its average value for our sample. Our simple calculation shows that for $P > P_{\mathrm{th}} = 2\pi G f(m) (\Delta V/2)^{-3}$, only systems with $e > e_{\mathrm{th}}$ would be detected, the eccentricity threshold value rising with P. This fact could thus have explained an $e - \log P$ correlation. In practice, however, $P_{\mathrm{th}} \simeq 1000$ d, too large a value to explain the observed correlation for most of our stars.

When $\Delta t \gg P$, we resort to the following procedure. First, we calculate the probability density of measuring, at given e, a radial velocity V (or more precisely F). Then we evaluate the probability density of the difference in V (i.e. in F) between two observations, assuming that both observations are independent (which is supposedly valid when $\Delta t \gg P$). This probability density is symmetric and centered round 0, its width being approximately proportional to the average of the absolute value of the radial velocity difference between two observations. One is therefore principally interested in the variance of this distribution:

$$\sigma^2 = \langle |\Delta F|^2 \rangle = \frac{2 (1 - e^2)^{3/2}}{e^2} \left[1 - \sqrt{1 - e^2} \right] \simeq 2 (1 - e^2)^{3/2} \qquad \text{for } e \sim 1 \quad . \tag{5}$$

The shape of the distribution clearly shows there is only a weak dependence on e: σ^2 only decreases significantly for e close to 1. We conclude that the observational bias cannot explain the lack of systems with $e \gtrsim 0.1$ in the short period region.

2.3.3.   Long period systems

Another striking point concerns the existence in our sample of long period systems (say $P > 130$ d) with small eccentricities ($e < 0.1$). Indeed, no systems with $P > P_{\mathrm{cut-off}}$ and $e < 0.1$ appear in the G-dwarfs sample of [DM91]. Likewise, the sample of open clusters red giants of Mermilliod and Mayor (this volume) clearly lacks systems with $e < 0.1$ and $P > 130$ d (only two stars out of more than 80 fall in that area). Although no convincing explanation has been proposed as to why such systems might not exist (do systems with initial $e < 0.1$ never form, or is there a mechanism increasing e when P exceeds $P_{cut-off}$?), we should ask ourselves why do such systems exist in our sample? We suggest an answer based on a comparison with Ba stars. Indeed, as will be shown in the next section, Ba stars have eccentricities smaller than those of normal giants. This supposedly results from a mass loss episode that turns an AGB star into a white dwarf. We may therefore well imagine those systems with large P and small e to contain a giant and a white dwarf. We are confident in this idea because the mass function of all but three systems pertains to the $0.002 - 0.05$ $M_\odot$ range, i.e. they all belong to the lower part of the normal giants $f(m)$ distribution (see Sect. 2.4), closely resembling Ba stars. One noticeable exception is HD 45910 ($\equiv$ AX Mon), a star known to experience a case B mass exchange [PCW91]. A very small eccentricity is expected in that case. The other exceptions, HD 2343 and HD 13738, have $f(m)$ of 0.193 and 0.218 $M_\odot$ respectively, which would require more massive white dwarfs ($\sim 0.8 - 1.2$ $M_\odot$). Another fact seemingly favouring a post-AGB phase for long P and small e stars is the existence of S1221 in M67 [MLG90]. This red giant binary, characterized by $P = 6450$ d and an almost circular orbit, presents a CN-enriched surface composition that might result from a (wind driven) mass transfer episode.

2.3.4. Barium stars

In Fig. 3 we display the $\langle e \rangle$ for our sample of Late-Type Giant Binaries, together with the data for Ba stars ([McCW90]; [JM92]). The figure shows that, at a given period, the eccentricity of Ba stars is generally smaller than the $\langle e \rangle$ eccentricity of normal field giants. Webbink [Web86] already drew that conclusion from a comparison of the eccentricity distribution of a small sample of red giants with that of the very low (at the time) number of Ba stars with known orbit. Several reasons may explain the smaller eccentricities (Jorissen & Boffin, this volume). We especially single one out.

The likely scenario leading to the formation of Ba stars [BJ88] implies that a huge mass loss took place in the binary system prior to the appearance of the Ba star. The mass loss leads to a considerable increase of the orbital period, while the eccentricity does not necessarily change very much. Therefore, the loci of Ba stars in the $e - \log P$ diagram should be shifted to larger P compared to field giants, and we tentatively fit their $\langle e \rangle$ with the relation

$$\langle e \rangle = 0.15 \left[(\log P - \eta) - 1 \right] \quad , \tag{6}$$

where η represents the magnitude of the shift. The very crude model of wind accretion outlined in [BJ88] predicts values of η in the range 0.3 to about 1, which are compatible with the rather small number of Ba stars with known orbital elements.

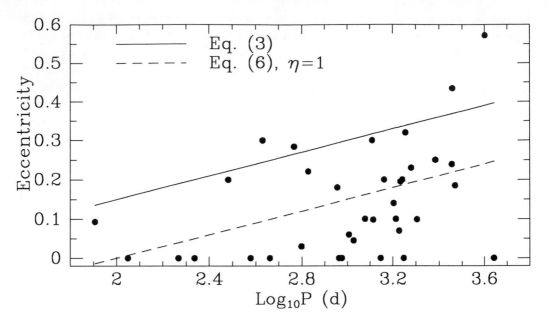

Figure 3. Eccentricity$-\log P$ diagram for Ba stars [McCW90]; [JM92]). The solid line represents the fit to the SB1 distribution, while the dashed line illustrates the same fit shifted by $\eta = 1$. The latter case applies to Ba stars.

2.4.   Mass function distribution

In order to extract the mass ratio distribution from the mass function one, we analyze the latter distribution by means of two different approaches, the Richardson-Lucy iterative inversion procedure on one hand and the Monte Carlo method on the other hand.

Devised first by Richardson [Ric72], the iterative inversion technique was independently rediscovered by Lucy [Luc74] in the framework of astronomical inverse problems. The method relies on the Bayes theorem on conditional probabilities, and solves the integral equation resulting from the convolution by an iterative scheme. Applying this procedure to our problem requires to assume the constancy of the giant mass m_G, which should be considered as a first order approximation. Let us now consider the distribution of $Y = f(m)/m_G$ (or more precisely that of $\log Y$, as $f(m)$ spans more than four decades). Inverting the integral equation

$$\Phi(Y) = \int \Psi(q) \, \Pi(Y \mid q) \, \mathrm{d}q \qquad (7)$$

gives the mass ratio distribution (MRD) $\Psi(q)$ we seek. The observed distribution of Y is $\Phi(Y)$ and the kernel $\Pi(Y \mid q)$ represents the conditional probability density to observe Y when the mass ratio q is known.

Using the definition of $f(m)$ and assuming that i distributes as $\sin i$, we calculate the kernel as follows. Writing

$$Y \equiv \frac{f(m)}{m_G} = \frac{q^3}{(1+q)^2} \sin^3 i = Q \sin^3 i \quad, \tag{8}$$

the probability density to observe Y for a certain value of Q (or, equivalently, q) is given by

$$\Pi(Y \mid q) = \int_0^{\pi/2} \delta(Q \sin^3 i - Y) \sin i \, di$$

$$= \left(3 \, Q^{1/3} Y^{1/3} \sqrt{Q^{2/3} - Y^{2/3}} \right)^{-1} \Theta(Q - Y) \quad, \tag{9}$$

where $\delta(x)$ and $\Theta(x)$ are the Dirac function and the Heaviside step function, respectively. The inversion of the integral equation is accomplished with the help of the Richardson-Lucy iterative method. We choose the number of iterations in such a way that the distribution $\Phi(Y)$ deduced from the calculated MRD does not differ too much from the observed distribution. However, small differences, ascribable to statistical errors in our sample of stars, are acceptable. Thus we do not have to reach full convergence of the method, for in that case both distributions would perfectly agree, which is undesirable (see Lucy [Luc74] for details). To our knowledge, it is the first time one applies the Richardson-Lucy technique to derive a mass ratio distribution from a mass function distribution. Figure 4 collects our results. We computed the MRD for two values of the primary's mass, namely 1.5 and 3 $M_\odot$. For $m_G = 1.5 \, M_\odot$, we obtain a smooth distribution with a small peak in the bin $q = 0.6 - 0.7$. In the case $m_G = 3 \, M_\odot$, we find a more pronounced peak in the range $q = 0.4 - 0.5$ and a lack of systems with $q > 0.7$.

As an alternate method to compute the MRD we performed a Monte Carlo simulation, adopting several trial distributions for the giant mass and the MRD and comparing the resulting MFD with the observed one. We will avoid here a lengthy discussion of all the distributions combinations that do or do not match the observed MFD. We limit ourselves to a presentation of our findings. First of all, it appears that a power law for the distribution of m_G does not give as good an accord with the observations as a constant m_G. Consequently, we have hereunder taken a constant value for m_G. Secondly, a distribution of q, uniform in the range $0.1 - 1$ and 0 elsewhere, produces the best fit for reasonable values of the giant mass, i.e. $m_G \simeq 1 - 1.5 \, M_\odot$. Thirdly, whatever the mass of the giant, a distribution of the kind $f(q) \propto q$, which favours high q values, disagrees with the observed distribution. Fourthly, a distribution that disfavours large mass ratios (like $f(q) \propto q^{-1}$) does not fit at all the observed

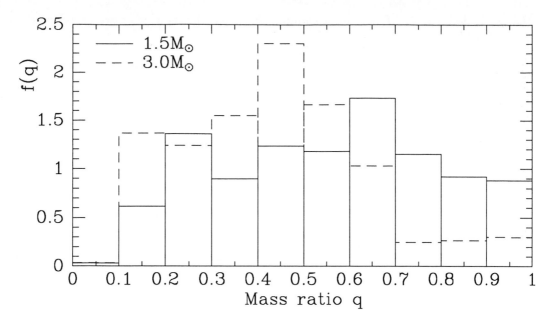

Figure 4. Mass ratio distribution derived from the distribution of mass functions of our catalogue, using the Richardson-Lucy method.

distribution. We especially checked that the solution found by Trimble [Tri90] gives a MFD very far from the observed distribution. Fifthly, taking a distribution for the inclination angle proportional to $\sin^2 i$ instead of $\sin i$ hardly alters the above results. In Figure 5 we show an example of a good fit ($f(q) = 1$) and of a bad one ($f(q) \propto q^{-1}$), respectively.

Combining both ways of calculating the MRD, we conclude that (i) for all values of the mass of the primary, the Richardson-Lucy method leads to a distribution of q displaying a peak whose position and magnitude depend on m_G; (ii) with the Monte Carlo technique, it appears that for simple q distributions small primary masses m_G are preferred ($\lesssim 2\ M_\odot$), although we cannot really discard values of $m_G = 3\ M_\odot$. The best fit is obtained with a constant mass ratio distribution. In fact it seems that the distribution of q should be such as to allow companion masses $m_C \simeq 1\ M_\odot$ to be well represented in the sample.

2.5.   Distribution of the semi-major axis

The distance separating the two stars in spectroscopic binaries, as well as their respective masses, are unknown. However, we can find the distribution of the separation by using our previous results. Taking the observed period distribution, a constant giant mass $m_G = 1.5\ M_\odot$ and a constant MRD for q between 0.1 and 1, we run a Monte

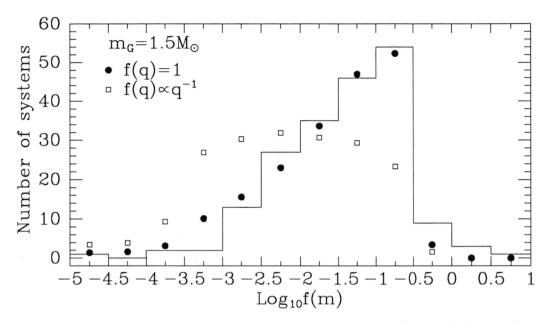

Figure 5. Distribution of $\log f(m)$ computed with the Monte Carlo technique using a giant mass $m_G = 1.5\ M_\odot$ and $f(q) = 1$ ($\bullet$) or $f(q) \propto q^{-1}$ ($\square$), compared to the observed one (solid line).

Carlo simulation to obtain the distribution of the logarithm of the semi-major axis. Figure 6 illustrates the calculated distribution. Its shape is that of a monotonically increasing function with a maximum at about $a \simeq 2.5$ AU. No systems are found with a separation below 0.02 AU, i.e. 4 $R_\odot$.

3.   Conclusions

The statistical analysis of the SB1 orbital elements lead us to the following conclusions. The period distribution, showing an abrupt decrease at $\log P \simeq 3.5$, compares well with that in [DM91] for field G-dwarfs. The eccentricity distribution features a substantial number of nearly circularised systems. The circularization cut-off period we find ($\log P_{\text{cut-off}} \simeq 1.7$) is $\sim \frac{1}{3}$ that of Mermilliod and Mayor (this volume). The fact we deal with field giants exhibiting a scatter in masses and ages might well account for the presence of systems with circular orbits up to periods on the order of 150 d. The long period ($\gtrsim 200$ d), small eccentricity (< 0.1) systems are interpreted as binary systems made of a giant and a white dwarf, situation resulting from a previous mass transfer episode. We have calculated the mass ratio distribution with the aid of two complementary methods: the Richardson-Lucy iterative inversion procedure, and a Monte Carlo simulation. The best agreement with the observed mass function distribution is obtained for an uniform MRD assuming an average giant

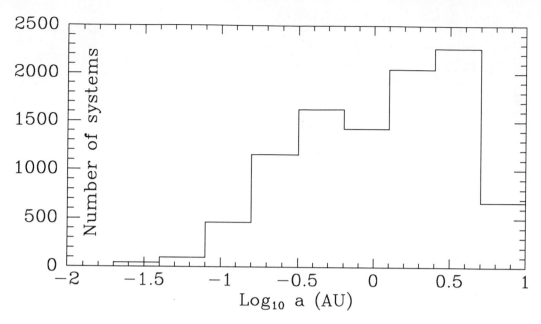

Figure 6. Computed distribution of the logarithm of the semi-major axis obtained through a Monte Carlo simulation.

mass $m_G = 1.5\ M_\odot$. This result contrasts with the conclusion of Trimble [Tri90] who claimed to find a good fit to the observed MFD with a MRD $f(q) \propto q^{-1}$ and a mean giant mass of $3\ M_\odot$. The semi-major axis distribution we calculate monotonically increases, presenting a maximum at 2.5 AU.

References

[BJ88]   Boffin, H.M.J., Jorissen, A., 1988, *A&A 205, 155*

[BCP92] Boffin, H.M.J., Cerf, N., Paulus, G., 1992, *A&A (submitted)*

[DM91]   Duquennoy, A., Mayor, M., 1991, *A&A 248, 485*

[JM92]   Jorissen, A., Mayor, M., 1992, *A&A (in press)*

[Luc74]   Lucy, L., 1974, *AJ 79, 745*

[MLG90] Mathieu, R.D., Latham, D.W., Griffin, R.F., 1990, *AJ 100, 1859*

[McCW90] Mc Clure, R.D., Woodsworth, A.W., 1990, *ApJ 352, 709*

[PCW91] Pols, O.R., Cote, J., Waters, L.B.F.M., Heise, J., 1991, *A&A 241, 419*

[Ric72]   Richardson, W.H., 1972, *J. Optical Soc. America 62, 55*

[Tri90]   Trimble, V., 1990, *MNRAS 242, 79*

[Web86]   Webbink, R.F., 1986, *in: Critical Observations versus Physical Models for Close Binary Systems, eds. K.-C. Leung and D.S. Zhai, Gordon and Breach, New York, p. 403*

[Zah77]   Zahn, J.P., 1977, *A&A 57, 383*

The Formation of Binary Stars

Cathy J. Clarke *

1. Introduction

It is now twenty years since modern computing power was first brought to bear on the problem of binary star formation (see the seminal work of Larson [Lar72] for a description of early work in this area). In this review, we shall concentrate on the progress of the last two decades and attempt to summarise the current status of a variety of theoretical models. For a more historical perspective, the reader is directed to Pringle [Pri91]; recent reviews are also given by Bodenheimer *et al.* [BRM91] and Boss [Bos91a]).

Binary star formation theories may, of course, be categorised in a variety of ways, but perhaps the most fundamental distinction is between those in which stars form singly within a cluster and subsequently pair up, and those in which stars form as binaries as a result of a splitting into two during the star formation process. In the former (capture) class of models, the theory of single star formation need be in no way modified and the problem becomes one of finding a mechanism by which two stars can lose some part of their energy of relative motion, thereby being deposited in a bound orbit. In the latter (fission/fragmentation) class of models, the binary fragments are always bound to one another and the problem then becomes one of achieving the necessary split into two at some stage of the star formation process. In Figure 1 we schematically illustrate the chief categories of star formation models. Amongst capture models, we distinguish between the case of many body and few body clusters. Amongst fragmentation/fission theories, models are classified according to the stage of the star formation process at which the split occurs: viz prior to the collapse of the parent condensations of molecular gas, during the collapse, following the collapse into an angular momentum supported disc and, finally, once the collapse has proceeded as far as the formation of a rapidly rotating protostar. (It should be stressed that these categories are more conceptual than physical, since each of these stages overlaps during the formation of a single star.)

In Section 2, we review each of these scenarios, concentrating in particular on the feasibility of the physical processes involved. Demonstrating feasibility is only a

*Institute of Astronomy, Madingley Road, Cambridge CB3 OHA, England.

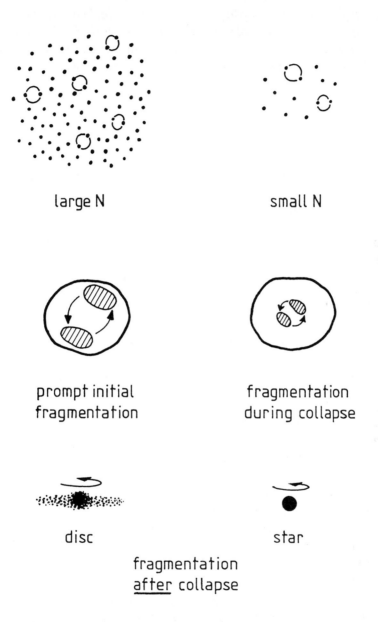

large N   small N

prompt initial   fragmentation
fragmentation   during collapse

disc   star

fragmentation
<u>after</u> collapse

Figure 1. Theoretical Models for Binary Star Formation.

partial goal for a successful model, however, since it should also be able to reproduce a wide range of observational parameters for main-sequence systems (see the Chapter by Duquennoy, this volume). In this respect, we argue in Section 3, theoretical models fall lamentably short of their goals. We suggest that the difficulty in mapping models into a precise set of observational predictions results largely from an ignorance of the effects of orbital evolution during the pre-main sequence stage. We therefore argue that the predictive power of theoretical models is severely limited by the existence of such a 'theory gap'. It is clear that recent and future observations of pre-main sequence (PMS) binaries will play a crucial role in closing this gap (see Chapters by Mathieu and Zinnecker, this volume).

2.   Binary formation models: physical feasibility

2.1.   Capture Theories

2.1.1.   Large N Clusters

The formation of a binary from two initially unbound stars requires the removal of energy from the relative orbit of the pair. One way in which this can be achieved (even in the absence of any dissipative processes) is if this energy is carried off in the form of kinetic energy by a third star in the cluster. This three-body capture process however requires that the three stars be simultaneously concentrated in a sphere of radius $\sim GM/v^2$ (for stellar mass M and cluster velocity dispersion v). For a cluster with v of ~ 1 km/s, this radius is ~ 1000 A.U., so that such capture events are clearly rare. Even in the dense environments of globular cluster cores it may be shown that this process is a negligible source of binaries [Hil76].

We turn, therefore, to dissipative capture processes. One source of dissipation is through the raising of tidal distortions in stellar envelopes during close star-star passages. This process is a likely source of Low Mass X-ray Binaries in the cores of globular clusters, where the two stars involved are respectively a low mass dwarf and a neutron star [FPR75]. However, since tidal capture is effective only for encounters in which pericentre is a few stellar radii, it is clear that this process can provide only the closest binaries, and even then, only in the densest environments.

However, during the pre-main sequence stage, it is possible that dissipative processes may be effective at larger radii, owing to the presence of extended discs of circumstellar gas and dust (for a review of the properties of protostellar discs see, for example, Becwith [Bec90]). In this case, capture might be effected through the action of star-disc drag, resulting from the transit of a star close to, or through, the disc around a neighbouring star. Larson [Lar90] argued that this process would be a significant source of binaries, although this conclusion required rather extreme assumptions concerning disc masses and radii, the efficacy of the capture process and the stellar densities in star forming regions. A further problem, which can reduce capture rates significantly, is that high velocity stars in the cluster can destroy the discs through which they pass, without themselves being captured [CP91a]. Although a definitive conclusion awaits a proper treatment of the star-disc drag process, such arguments suggest that such captures may not be a major source of binaries, at least, not in the high velocity environments of large N, virialised clusters. If, however, fragmentation and star formation were to occur on a timescale less than or comparable to the dynamical timescale of the parent cloud (as has been argued in models of globular cluster formation, e.g. [ML89]), then the stars are formed with velocities that are substantially sub-virial. In this case, a binary fraction of up to 10 per cent may be generated by star-disc captures during the initial collapse of the cluster [MCP91].

2.1.2. Small N clusters

In few-body systems, the gravitational interactions between neighbouring stars are relatively important, compared with large N systems, because each star provides a significant fraction of the total potential of the system. Consequently, the two-body relaxation time is relatively short. In a series of numerical simulations, van Albada [VA68a], [VA68b], showed that small N systems dissolve as a result of two body relaxation within a timescale of order N crossing times (this timescale is somewhat dependent on the mass spectrum employed). The energy required to dissolve the cluster is released by the (dynamical) formation of (usually one) binary. Therefore, in small N clusters, purely dynamical interactions are a viable source of binaries. However, since only one binary is formed per cluster, it is clear that the binary formation rate is too low to reproduce binary statistics (unless, of course, each mini-cluster is of extremely small N (two or three), in which case the problem then becomes one of explaining the fragmentation process, see below). The situation may be improved if, in addition to dynamical processes, some dissipative process (such as star-disc interaction) also plays a role [CP91b]; the star-disc capture mechanism is likely to be more effective in the small N environment, both because it is assisted by dynamical processes and because the lower velocity dispersion reduces the destruction of discs by fast star-disc impacts.

2.2. Fragmentation/Fission Models

2.2.1. Prompt Initial Fragmentation

In this scenario, the split into two occurs prior to the collapse of a bound condensation of gas to stellar dimensions. Such condensations of molecular gas may be observationally identifiable as the dense cores detected in NH_3 (Myers and Benson [MB83]). The statistics of such cores, relative to the rate of local star formation, implies that they are not collapsing on a free-fall time: the star formation process therefore 'hangs up' at a size scale of $\sim$ 10000 A.U.. At this scale, magnetic fields may provide the dominant support, so that further collapse requires the removal of magnetic flux by some process. We here consider the case where this support is removed rapidly (i.e. on less than a dynamical timescale), leaving a configuration that is highly gravitationally unstable, and where the removal of pressure support (e.g. through a core-core collision) imparts a large (order unity) deviation from axisymmetry. This situation should be contrasted with that considered in Section 2.2.2. below, in which case collapse proceeds from almost axisymmetric initial conditions.

One possibility, explored in recent work by the Cardiff group (Pongracic *et al.* [PCD92]) is that core collapse is initiated by the collision between two cores. For typical core velocity dispersions of 1-2 km/s, such collisions are both highly supersonic and highly super-Alfvenic and therefore strongly dissipative. These calculations are incomplete in the sense that they omit magnetic fields, but give a useful indication of the physics involved in the limit that the field is dynamically unimportant (i.e. they apply either to the case in which the cores were primarily thermally supported, prior to the collision, or else to the case where the collision effectively removes the field via magnetic reconnection). The striking result of this calculation is that the collision effects a collapse to near protostellar densities (i.e. a density enhancement of factor 10^{10}) on a timescale less than the dynamical timescale of the original core. This effect may be traced to the strong positive density dependence of optically thin cooling processes in the temperature range 10-100K: thus the dense shocked gas at the core- core interface cools efficiently, promoting gravitational instability and hence further compression, leading in its turn to further cooling, and so on. For the purpose of this discussion, however, the most interesting effect is that whereas head on-collisions generate a single bound condensation ('star'), in grazing collisions two such stars are produced, some of the angular momentum of the relative orbit of the two cores being transferred to the orbit of a stellar binary [PCD92]. Whether, of course, this binary is bound depends on the details of the collision: the impact parameter, the details of angular momentum transfer during the collision and subsequent collapse. Further simulations are required to determine over what range of collision parameters are (bound) binary systems produced, and thus whether this mechanism could contribute significantly to binary statistics.

Another recent simulation involving binary formation in the context of highly non-equilibrium conditions is presented by Bonnel *et al.* [BMB91]. In this case the initial condition is that of a prolate molecular core, a geometry favoured by Myers *et al.* [MFG91] in their interpretation of the statistics of observed aspect ratios of molecular cores. It is well known, however, (e.g. [Mes65]) that prolate configurations supported by purely thermal pressure are unstable against collapse; presumably. therefore, some anisotropic support (probably magnetic in origin, [Shu91]) is required to stabilise such cores, prior to their collapse. The calculation of [BMB91] again omits magnetic fields and therefore implicitly assumes that such support has been removed on a dynamical timescale, leaving a configuration that is highly dynamically unstable. This then collapses to a spindle, simultaneously breaking into two (or more) fragments along its length, a result also found in previous grid based calculations [Lar72]. These fragments fall together along the spindle axis, but avoid re-coalescence if the initial core possesses finite angular momentum normal to its long axis (i.e. if it rotates end over end). The result, in this case, is a highly eccentric binary, whose pericentre depends on the core's rotation and whose major axis is initially comparable with the dimension of the core.

Clearly, each of these pictures is highly incomplete, in greatest part because neither addresses in any detail the loss of magnetic support that must precede the events described above. The results of these simulations however point to a significant difference between such models and those which start from nearly equilibrium initial conditions, namely that, after a dynamical timescale or so, the mass incorporated in the proto-binary structure is significant compared with the mass bound to the proto-binary: in the model of [BMB91], for example, between one half and two thirds of the cloud mass ends up in binary (or higher order multiple) fragments. This may be regarded as an attractive feature of such models, since it reduces the risk that the binary will be destroyed by transferring its angular momentum to outward lying material. The relatively high efficiency of binary formation in this context may be largely ascribed to the large deviations from equilibrium in the initial conditions; in addition, in core-core collision models, the high Mach number of the impact results in a large fraction of the residual core material becoming unbound with respect to the binary.

2.2.2. Fragmentation during Collapse

This question has been the subject of more investigation than any other in the realm of binary star formation (see, for example, [Lar78], [BTB80], [Bos86], [Bos91b] and references therein). In each case, the initial configuration is a rotating cloud containing a number (greater than one) of Jeans masses. Non-axisymmetric perturbations are applied to this cloud and the collapse is then followed for as long as is feasible with grid based hydrodynamic schemes (that is, for one to two initial free-fall timescales). In the limit that the perturbations are of large amplitude (i.e. order unity) and that the number of Jeans masses in the cloud is large, this approach tends to that described in Section 2.2.1 above. However, the motivation behind the work described in this Section has generally been to discover whether binary fragmentation is a natural outcome of mildly non-axisymmetric collapse, and therefore the approach has been to keep the initial perturbations as small as possible.

To summarise briefly the results of the aforementioned experiments, it can be said that binary fragmentation can often be induced by the imposition of quite modest (of order ten percent) m=2 perturbations *prior* to the collapse. It is notable, however, that non-axisymmetric structure is not amplified by a large factor during the collapse, i.e. ten percent perturbations give rise to proto-binaries which, at the end of the numerical experiment, contain of order ten percent of the system mass. Several conditions must be met, however, if such structure is not to be damped out by pressure and tidal forces during the collapse. One such constraint concerns the initial density profile in the cloud: even a rather moderate degree of central concentration (i.e. a r^{-1} density profile) is sufficient to inhibit binary fragmentation [Bos86]. This implies, therefore, that the dynamical collapse treated in such calculations *cannot* have been preceded by a phase of quasi-static contraction (as, for example, in models in which magnetic support is lost slowly through ambipolar diffusion), since such a phase leads to the establishment of a r^{-2} density profile [LS89] - too steep to permit binary fragmentation. Secondly, the survival of proto-binary structure also requires that the collapse is halted by centrifugal forces before the optical depth in the cloud becomes greater than unity, i.e. whilst the collapse is still approximately isothermal. In the case of more slowly rotating cores, the collapse enters the adiabatic regime, whereupon pressure forces become efficient in erasing non-axisymmetric structure. The transition between optically thin and optically thick regimes occurs at about 100 A.U., implying that binaries wider than this may be formed rather readily by retention of non-axisymmetric structure in the pre-collapse core. In the case of closer binaries, the survival of such structure is much more problematical, even over a dynamical timescale, and requires an extremely low thermal energy content in the initial cloud [Bos86].

The unanswered question that lingers in the case of all such calculations is however the fate of the proto-binary on longer timescales, since grid-based hydrodynamic calculations are able to pursue the collapse for only one to two initial free-fall times of the core. As noted above, the fraction of the core mass contained in the proto-binary at this stage is small, particularly in the case of the closer systems generated through multiple fragmentations. It is yet to be demonstrated that, under these circumstances, the proto-binary can survive the pre-main sequence stage for it is not at all clear what the subsequent evolution would be: i.e. whether the binary would survive, acting as a seed onto which the residual gas accretes, or whether, conversely, it would lose angular momentum to the residual gas, thus spiralling in and re-coalescing to form a central star and extended disc. It is likely that this question will remain unanswered until numerical investigations have been pursued to the point at which the bulk of the material is incorporated in the proto-binary. Such an aim is however beyond the scope of grid-based hydrodynamic methods, owing to the large dynamic range of timescales and densities in the problem but may be feasible with particle-based methods. It should be noted that this problem of determining the ultimate fate of binaries that are bound to significant quantities of circumbinary gas is one which is encountered, at some level, in all binary formation scenarios. It is particularly acute in the present scenario, however, because of the modest deviations from axisymmetry

7

in the initial conditions, and the consequently slow growth of proto-binary structure.

2.2.3.   Fragmentation after Collapse

We now turn to models in which binary fragmentation occurs after the collapse of the core. Early numerical work by Larson [Lar72] indicated that, if one suppresses non-axisymmetric instabilities, a rotating cloud collapses to form an angular momentum supported ring, fueling speculation that the break up of such a ring might give rise to the formation of a binary. Indeed, Norman and Wilson [NW78] demonstrated that if one applies non-axisymmetric perturbations to such a ring, it breaks up into a number of pieces, the number depending on the value of the azimuthal wavenumber, m, for the perturbation. This approach has been widely disputed however (Bodenheimer *et al.* [BTB80]), since it requires the suppression of all such instabilities during the collapse, and then their convenient reactivation once the bulk of the mass of the core has fallen into the ring. Self-consistency demands, instead, that the perturbations be applied prior to the collapse and be allowed to develop as the collapse proceeds: the approach employed, in fact, in the calculations described in Section 2.2.2 above.

The question has been re-opened, however, by recent work by Adams *et al.* [ARS89], in which they demonstrated, through linear stability analysis, that moderately massive discs surrounding a central protostar can be gravitationally unstable to an m=1 mode: such a mode had been neglected by previous workers since, in contrast to modes of higher order, it requires that the central star be allowed to move, relative to the centre of mass of the star-disc system. In this and a following paper (Shu *et al.* [STAR90]), it was argued that the wide spacing of the inner and outer Lindblad resonances for the m=1 mode in a nearly Keplerian disc would allow this mode to be global, thus giving rise to the hope that a significant fraction of the disc's mass would be involved. It was therefore envisaged that the outcome of the development of such a mode would be a binary, of which one member would be the original central star and the other would have been gathered up out of disc material. This is clearly an attractive scenario, given the widespread occurence of protostellar discs, but cannot be regarded as more than provisional pending full non-linear calculations. Amongst the features that must be demonstrated by such calculations is that the m=1 mode dominates over higher order modes, that a substantial fraction of the disc mass is involved in the instability and that, crucially, a threshold effect delays the onset of the instability until a substantial fraction of the mass of the parent core has fallen into the disc. If this final criterion is not met, this model is subject to the same objection as the fragmenting ring model described above.

Finally, we turn to a model with a long history of popularity, particularly for the formation of the closest binaries, that of the fission of a rapidly rotating protostar. Such models rely on the well known dynamical bar mode instability of a rotating spheroid, which sets in once the ratio of its rotational energy to gravitational energy exceeds a critical value (e.g. [Chr69]). The onset of such an instability, in the context of star formation, then requires that the protostar has retained a significant angular momentum during its formation process, or rather, in the parlance of modern star formation theories, that the protostar has been spun up to near break up by the accretion of high angular momentum material from its protostellar disc. Again, the success of such a model requires that the onset of instability be delayed until a significant fraction of the parent core has had time to accrete onto the central protostar.

However, even if one assumes that the requisite initial conditions for the instability can be set up during the collapse process, there remains the further problem of what is the likely outcome of the instability. In the case of an incompressible fluid, the result is clear - a rapidly rotating raindrop indeed splits into two, its initial angular momentum being partially transferred into the orbital angular momentum of the resultant pair. For a compressible fluid, the outcome is less clear on analytical grounds. Jeans [Jea29] suggested that in this case, instead of fission, the excess angular momentum might be removed via the ejection of a pair of spiral arms, leaving behind a single star, now rotating below the instability threshold. This conclusion was sufficiently controversial (see e.g. [Ost70]), that the problem was finally investigated numerically (Durissen *et al.* [DGTB86]). These authors tackled the problem with three independent codes (two grid-based- due to Boss and Tohline- and one SPH, due to Gingold). In each case it was found, in line with the speculations of Jeans, that the outward transfer of angular momentum by gravitational torques produced a central remnant (containing most of the mass) and a disc/ring of debris shed from the stellar equator (containing most of the angular momentum). These calculations appeared to demonstrate convincingly, therefore, that the outcome of such dynamical instability is *not* the formation of a binary.

3.   Observational predictions

In the previous Section, the discussion was centered on the physical feasibility of each of the suggested mechanisms for forming initial proto-binary structure. In this Section, we turn to the vexed question of the relationship of each of these models to the observed properties of binary systems. At this point, however, we encounter a grave problem in that essentially none of the models described above lead to the formation of binary structure in a 'clean' environment - that is, in all cases, the proto-binary is bound to a certain amount of residual gas, which can be distributed both around the individual binary components (circumstellar) and exterior to the binary orbit (circumbinary). Evidently, the larger the ratio of residual gas mass to binary mass, the more important is the subsequent dynamical evolution of the system, as energy and angular momentum are transferred between the binary and surrounding gas, and one may envisage that in the limit that this ratio becomes very large the very survival of the binary might be threatened. It might be argued, therefore, that such considerations might lead one to discriminate between contending models on the basis of the efficiency with which they incorporate gas into the proto-binary structure.

Even leaving aside the question of binary survival, it is likely that binary-gas interaction plays a crucial role in modifying the orbital parameters of binaries during the pre-main sequence stage: numerical simulations indicate that resonant coupling between a binary and a ring of circumbinary gas can cause significant orbital evolution even in the limit of a very modest gas to binary mass ratio (see Chapter by Lubow, this volume). It is likely, however, that the magnitude and sign of such changes depends on whether the residual gas is predominantly circumstellar ('interior disc') or circumbinary ('exterior disc'); moreover, in the case of predominantly resonant coupling, the orbital evolution is also a sensitive function of exactly which resonances are populated, itself a function of binary mass ratio. Until such questions are settled, therefore, a 'theory gap' separates the calculations described above from the set of observational parameters that they are designed to explain. In the light of this gap, therefore, all the remarks made below concerning the observational predictions of various models must be considered, at best, highly speculative.

Bearing in mind the caveat expressed in the preceding paragraph, what might one say about the properties of binaries produced by each of the mechanisms outlined above? In the case of capture models (Section 2.1), we have argued that the only feasible scenario for all but the closest binaries is that of star-disc capture: in this case the capture products initially form highly eccentric binaries, with a distribution of pericentres that peaks near the outer radius of the discs concerned. Evidently, therefore, in the absence of subsequent orbital evolution, such a scenario is incapable of explaining the entire distribution of binary periods, since it predicts a deficit of binaries of short periods, as well as failing to produce binaries wider than the outer disc radius. Substantial orbital evolution is required if close binaries are to be explained by this mechanism, involving an energy loss that is large compared to the initial binding energy of the capture products, as well as a significant damping of eccentricity. It is not clear whether binary-gas interaction can achieve the necessary evolution, however, since the sign of the transfer of energy and angular momentum between gas and binary depends on the distribution of bound gas following the capture event. Numerical simulations of encounters between stars and protostellar discs indicate that the residual gas is retained as circumstellar, rather than circumbinary, material [CP92]. In this case, the disc gas rotates with a higher angular velocity than the binary, so that energy may be transferred from the gas into the binary orbit, implying, therefore, a consequent *growth* of binary period. This lack of bound circumbinary material, following the capture event, is also notable in that it implies a possible observational discriminant between competing models: a systematic investigation of the spatial relationship between pre-main sequence binaries and surrounding emission would therefore be particularly timely.

We turn now to prompt initial fragmentation models (Section 2.2.1). Here the initial binary major axis is of order the radius of the parent core (i.e. 0.01 to 0.1 pc), and such orbits are, as above, initially highly eccentric. Two features distinguish this case from the star-disc capture scenario, however. Firstly, the pericentre is determined not by the dimension of a pre-existing disc, but by the angular momentum of the parent core, with slowly rotating cores producing more radial orbits. However, the initial radius of the discs formed around the individual stars, following the onset of collapse, is also determined by the angular momentum of the core; as a result, therefore, the discs around the individual stars always intersect at binary pericentre, irrespective of the core rotation [Pri89]. This raises the possibility of substantial dissipation of orbital energy of the binary, due to star-gas drag, during the first pericentre passage. Secondly, it is not clear what predictions such models make for the distribution of circumbinary gas. This probably depends on the details of the fragmentation process involved, with collisionally induced fragmentation being relatively efficient in dispersing circumbinary material.

If, however, binaries form through fragmentation, either during the collapse or following the collapse to a disc (Section 2.2.2 and 2.2.3), then it is likely that they are formed with a substantial quantity of bound gas exterior to the initial binary orbit. This circumbinary material may play an important role in generating the significant eccentricities observed in main sequence binaries. Simulations indicate that the orbital evolution may be rapid in this case (Artymowicz *et al.* [ACL91]), and possibly even detectable (through precession of the line of apsides) in the case of an eccentric binary embedded in a circumbinary disc. (It is unfortunate, in this respect that the only pre-main sequence spectroscopic binary where there is evidence for a circumbinary disc should be in a precisely circular orbit (see Chapter by Mathieu, this volume).It is not at all clear, however, how a picture in which binaries are formed on circular orbits, and driven eccentric by interaction with circumbinary gas, can fit in with the observed increase of mean binary eccentricity with period (see Chapter by Duquennoy), since one would anticipate that such a driving mechanism would be more effective in the case of closer binaries.

Finally, having considered in turn the possible predictions that the more plausible models might make for the periods and eccentricities of main sequence binaries, and also for the gas distributions around pre-main sequence binaries, we turn briefly to the question of binary masses, mass ratios and binary frequency. Surprisingly little can be said on this point, however, in large part because the class of models that have been investigated most thoroughly (that is, fragmentation models) tend to use initial conditions that are an m=2 perturbation on an axisymmetric distribution, and which, inevitably therefore, produce binaries of unit mass ratio. It is notable, however, that the imposition of m=1 perturbations favours the production of much more extreme mass ratios [Bos90]. In the case of purely dynamical capture models there is a marked tendency for the two most massive stars in a cluster to pair up [VA68b]. This would imply a strong bias towards unit mass ratios (in contrast to the observed properties of low mass binaries) and also a systematic increase of binary frequency with stellar mass; it is unclear whether the inclusion of star-disc dissipation would change the mass dependence of the capture process, however. Finally it should also be noted that collisional fragmentation models also appear to favour an increase of binary frequency with stellar mass: nearly head-on collisions between cores tend to form more massive stars, in bound orbits, whereas grazing collisions produce lower mass stars in unbound orbits (Pongracic, private communication).

4.   Conclusions

We have reviewed a number of theoretical pictures of binary formation, from which it can be concluded that whereas certain models appear to be unlikely on theoretical grounds (e.g. large N dynamical capture or stellar fission), no clear front-runner has emerged from the remaining models. A number of scenarios hold considerable promise, however, and an obvious task for theorists is to further investigate the physical feasibility of such models. An equally pressing goal, and one that has received scant attention until recently, is that of closing the 'theory gap' concerning the orbital evolution of pre-main sequence binaries, due to interaction with residual gas. We have argued above that the uncertainties currently surrounding this process make it very hard to map a theoretical model onto a set of observational predictions - or even, in fact, to predict whether a given proto-binary can survive the pre-main sequence stage at all!

As far as tasks for observers are concerned, two areas of investigation would appear particularly fruitful in unravelling the problem of binary formation. One concerns an extension of the superb data sets now available for main sequence binaries with late F/G primaries, to other mass ranges. In particular, one would wish to know how the incidence of binarity changes as a function of system mass; another important question concerns at what mass is there a transition in the pairing properties of binaries, from the random association by mass that typifies G star binaries, to the pronounced bias towards unit mass ratios apparent in O star binaries (Garmany et al 1980 [GCM80]). Secondly, the study of pre-main sequence binaries, their orbital parameters and associated gas is likely to provide a crucial discriminant between contending models. Of particular interest is whether there are any systematic differences in the period/eccentricity distributions of main sequence and pre-main sequence binaries, since any such differences would point to the initial conditions of binary formation. As stressed in Section 3 above, the distribution of residual gas in the vicinity of proto-binaries can also provide an important clue as to the mechanism for binary formation: for example, little circumbinary material should be associated with binaries that result from star-disc capture, whereas in disc fragmentation models, considerable reservoirs of circumbinary material appear inevitable. Such preliminary conclusions should spur the pursuit of such observational programs, although further theoretical work is required before all the available information can be extracted from such data.

References

[ARS89] Adams, F.C., Ruden, S.P. and Shu, F.H. 1989. *Astrophys. J.* **347, 959.**

[ACL91] Artymowicz, P., Clarke, C.J., Lubow, S.H. and Pringle, J.E.. 1991. *Astrophys. J. Lett.* **370, L35.**

[Bec90] Beckwith, S. 1990. in *'Structure and Emission Properties of Accretion Discs'*, eds C. Bertout, S. Collin-Souffrin and J.-P. Lasota, Editions Frontières, p. 33.

[BMB91] Bonnell, I., Martel, H., Bastien, P., Arcoragi, J.-P. and Benz, W. 1991. *Astrophys. J. 377, 553.*

[BTB80] Bodenheimer, P., Tohline, J.E. and Black, D.C. 1980. *Astrophys. J. 242, 209.*

[BRM91] Bodenheimer, P., Ruzmaikhina, T. and Mathieu, R.D., 1991 in 'Protostars and Planets III' (eds Levy and Mathews), in press.

[Bos86] Boss, A.P., 1986. *Astrophys. J. Suppl. 62, 519.*

[Bos90] Boss, A.P., 1990. in 'Protostars and Planets III', (eds Levy and Mathews) in press.

[Bos91a] Boss, A.P., 1991a) in 'Interacting Binary Stars', J. Sahade, G. McCluskey and Y. Kondo (eds), Kluwer, Dordrecht.

[Bos91b] Boss, A.P., 1991b). *Nature 351, 298.*

[Chr69] Chandrasekhar, 1969 in *'Ellipsoidal Figures of Equilibrium'*.

[CP91a] Clarke, C.J. and Pringle, J.E., 1991a). *M.N.R.A.S. 249, 584.*

[CP91b] Clarke, C.J. and Pringle, J.E., 1991b). *M.N.R.A.S. 249, 588.*

[CP92] Clarke, C.J. and Pringle, J.E., 1992 in preparation.

[DGTB86] Durisen, R.H., Gingold, R.A., Tohline, J.E. and Boss, A.P., 1986. *Astrophys. J. 305, 281.*

[FPR75] Fabian, A. C., Pringle, J.E. and Rees, M.J., 1975. *M.N.R.A.S 172, 15P.*

[GCM80] Garmany, C., Conti, P.S. and Massey, P., 1980.*Astrophys. J. 242, 1063.*

[Hil76] Hills, J.G., 1976. *M.N.R.A.S. 175, 1P.*

[Jea29] Jeans, J. 1929. *'Astronomy and Cosmogony'*, 2nd Edition, C.U.P..

[Lar72] Larson, R.B., 1972. *M.N.R.A.S. 156, 437.*

[Lar78] Larson, R.B., 1978. *M.N.R.A.S. 184, 69.*

[Lar90] Larson, R.B., 1990. in *'Physical Processes in Fragmentation and Star Formation'*, R. Capuzzo-Dolcetta, C. Chiosi and A. di Fazio (eds), Kluwer, Dordrecht, p. 389.

[LS89] Lizano, S. and Shu, F.H. 1989. *Astrophys. J. 342, 834.*

[Mes65] Mestel, L. 1965. *Q. Jl. R. Astr. Soc. 6, 161.*

[MCP91] Murray, S.D., Clarke, C.J. and Pringle, J.E., 1991. *Astrophys. J. 383, 192.*

[ML89] Murray, S.D. and Lin, D.N.C., 1989. *Astrophys. J. 339, 933.*

[MFG91] Myers, P.C., Fuller, G., Goodman, A.A. and Benson, P.J. 1991. *Astrophys. J. 376, 561.*

[MB83] Myers, P.C. and Benson, P.J. 1983. *Astrophys. J. 266, 309.*

[NW78] Norman, M.L. and Wilson, J.R. 1978. *Astrophys. J. 224, 497.*

[Ost70] Ostriker, J.P. 1970 in *'Stellar Rotation'*, A. Shettebak (ed.) Gordon and Breach, New York,p. 147.

[PCD92] Pongracic, H., Chapman, S.J., Davies, J.R., Disney, M.J., Nelson, A.H., and Whitworth, A.P. 1992. *M.N.R.A.S. in press.*

[Pri89] Pringle, J.E. 1989. *M.N.R.A.S. 239, 361.*

[Pri91] Pringle, J.E. 1991 in *'The Physics of Star Formation and Early Stellar Evolution'*, C. Lada and N. Kylafis (eds.), Kluwer, Dordrecht, p. 437.

[Shu91] Shu, F.H. 1991 in *'The Physics of Star Formation and Early Stellar Evolution'*, C. Lada and N. Kylafis (eds.), Kluwer, Dordrecht, p. 365.

[STAR90] Shu, F.H., Tremaine, S., Adams, F.C. and Ruden, S.P. 1990. *Astrophys. J. Lett. 370, L31.*

[VA68a] van Albada, T. 1968a). *Bull. Astr. Inst. Neth. 19, 479.*

[VA68b] van Albada, T. 1968b). *Bull. Astr. Inst. Neth. 20, 57.*

Distribution and Evolution of Orbital Elements for $1\mathcal{M}_\odot$ Primaries

Antoine Duquennoy and Michel Mayor [*]
Jean-Claude Mermilliod [†]

Abstract

Stellar duplicity surveys, through the distributions of orbital elements, are important in the fields of binary formation and of binary evolution processes. Here we present observational results currently being obtained from CORAVEL radial-velocity surveys on stellar duplicity among nearby stars and among open clusters.

We then explore numerically the variations of the orbital period and eccentricity under tidal effects, assumed to occur on the MS among late-type binaries according to Zahn's theory. We find that i) eccentric orbits with period shorter than the current definition of the cut-off period can be explained by incomplete tidal circularization, ii) a new definition of this cut-off period must be adopted, and we propose the period of the longest period circular orbit in a homogeneous sample.

We finally display the available data on all binaries of known age in a plane orbital period vs age, distinguishing the circular orbits from the eccentric ones. We conclude that this representation supports the above definition of the tidal cut-off period, but that the observational constraints on the tidal theories, in particular on the circularization timescales on the PMS and on the MS remain weak.

1. Introduction

The observation of duplicity in various stellar populations allows to study how do behave, maybe how do form, the binaries according to various parameters such as their mass, their birthdate (age), or their birthplace in the Galaxy (disc/halo), through the distributions of their orbital elements.

[*]Observatoire de Genève, 51 chemin des maillettes, CH 1290 Sauverny, Switzerland
[†]Institut d'Astronomie de l'Université de Lausanne, 51 chemin des maillettes, CH 1290 Chavannes-des-Bois, Switzerland

However, until a few years ago, all the studies of duplicity had to deal with huge observational biases such as limiting magnitude or timespan of the observations. These studies (even now) also suffer from generally too small numbers of binaries when one looks to the possible correlations between orbital elements. But more and more observations of binaries are now made in several observatories, thanks to new techniques. In particular in the spectroscopic binary (SB) domain (see e.g. the review by Latham [Lat85]), the apparition of radial velocity (RV) determination by cross-correlation [Gri67], allowed a great observational effort to derive the caracteristics of SBs, reducing the biases. We note, for example, that more stars have been observed during the last 13 years with the two CORAVEL spectrometers (described by Baranne et al. [BMP79]) than during the first century of classical spectroscopic RV determination.

In parallel, and as already reported by several authors (see [Zah89], [GoM91]), the recent years have seen a renewed interest in the study of the evolution of close binary systems, in particular under the action of tidal effects (synchronization, circularization of the orbits) between the two stellar components. These studies provide a key to understand the presently observed distributions of orbital elements in homogeneous samples (see e.g. analyses in [MaM84], [MLG90], [DuM91]), which in turn are potentially strong constraints on the mechanisms of binary formation ([Bos88], [Pri89], [Cla92]). Besides, the question of whether the orbital circularization mechanism can provide a new clock to date coeval samples of binaries has been suggested ([MaM84], [MaM88]) but the conclusions remain weak (this paper, and [MDL92] in this volume).

We will restrict this paper to essentially solar-mass spectroscopic binaries, with periods shorter than about 3000 days. We will first recall and present some results being obtained from long-term radial velocity surveys of stellar duplicity, in particular those conducted with the CORAVEL spectrometer. The accent will be put on the unbiased distributions of orbital elements in these samples.
In section 3 we will review some of the mechanisms of orbital element evolution and some of the theoretical approaches followed by various authors. Then we will examine in particular the case of tidal effects in short period ($P < 25d$) main-sequence binaries, using the prescriptions of Zahn [Zah77]. Simulations of the evolution of the orbital period P and of the eccentricity e assumed to occur during the main-sequence will be presented and we will derive some interesting results about the significance of the observed cut-off in orbital periods which will lead to redefine this cut-off.
In section 4 we display the existing data on the elements (P, e) for various samples of binaries with roughly solar mass primaries in the $(\log P)$ vs $(\log age)$ plane. We briefly recall the previous interpretations of the cut-off period and propose a new one in light of the approach made in section 3. A detailed discusion of the confrontation between tidal theories and observations is made by Mathieu et al. [MDL92] in this volume. We conclude this paper with some interesting prospects to come in the field of stellar duplicity.

2. The observed distributions of orbital elements among solar-like binaries

2.1. Among nearby G-dwarfs

Astronomers studying stellar duplicity are (or should be) now aware of the observational biases which can affect their sample. These biases have been reported and analyzed by many authors (e.g. [Hal87], [Tri90]).

The Gliese [Gli69] catalogue of nearby stars still provides a unique - and, as far as possible, complete - sample of stars with known distance, until the next achievement of the astrometric satellite HIPPARCOS mission. Thus it allows the study of well defined, unbiased samples of nearby field binaries that can be separated in three subsamples according to their primary masses (namely the G, K and M dwarfs). These stars are continuously surveyed with the CORAVEL spectrometer since almost 15 yrs. The radial-velocity precision ranges from 0.3 km/s for bright G-dwarfs to typically 1 km/s for faint ($m_v = 12$) M-dwarfs.

The duplicity in a complete sample of 164 nearby G-dwarfs primaries selected in the Gliese catalogue has been recently analysed by Duquennoy and Mayor [DuM91], using about 4200 CORAVEL radial-velocity measures and the existing data in the litterature for the visual binaries (VBs) and common proper motion (CPM) components. The selection biases are assumed to be negligible, the sample being parallax limited ($\pi_{trig} > 0.045$") instead of magnitude limited (see Branch effect, [Bra76]). So the authors only corrected the observed distributions of orbital elements (P, e, and the secondary mass M_2) from the detection biases due to instrumental limitations and to the observational procedure. These biases were corrected using numerical simulations for the SBs, and using Halbwachs [Hal87] method for the VBs and CPM components. We recall here the main results of [DuM91] on the corrected distributions directly related to the purpose of this workshop:

i) the orbital period distribution appears remarkably close to a gaussian curve, on a period range of not less than 11 decades, and with a median period of about 180 yrs. This last value is to be compared with the lower value of 14 yrs found by Abt and Levy [AbL76] in a sample of 132 G-dwarfs contained in the Bright Star Catalogue [Hof82]. The difference has been interpreted by [DuM91] in terms of the numerous short-P orbits in [AbL76] study which were later demonstrated to be spurious by Morbey and Griffin [MoG87].

ii) the orbital eccentricity distribution for orbital periods less than 3000 d shows clearly a cut-off around $P_{cut} = 11$d. Below this value the orbits have essentially zero eccentricities (say $e < 0.03$), at least within the computed errors. This is now commonly interpreted in terms of tidal effects acting between the two components of the system. The question to know when these effects principally take place, whether on the pre-main-sequence (PMS) stage or on the main sequence (MS), is one of the central debates of this workshop. Similar cut-off periods are found in other samples (see a discussion in Sec.4 in this paper, and by [MDL92] in this volume). Above the value of $P_{cut} = 11d$, the orbits of nearby G-dwarf stars all have eccentric orbits ($e > 0.03$). The eccentricity distribution is then bell-shaped with a mean value $\bar{e} = 0.31 \pm 0.04$ for 17 binaries. [DuM91] noted that one observes roughly the same mean eccentricity value for SBs with $P_{cut} < P < 3000d$ among at least two other samples: the galactic halo (orbital elements taken from [LMC88]), and a sample of binaries with probable very low mass companions, taken among CORAVEL SBs and astrometric binaries. They suggest that the mean eccentricity of primordial binaries (i.e. in that later range of periods, not affected by tidal effects , wind accretion nor larger scale interactions), is independent of the primary's age and of the secondary mass. We note here that this assertion is also verified for PMS stars: the 13 PMS binaries with $10 < P < 3000d$ quoted by Mathieu ([Mat89] and references therein) have $\bar{e} = 0.33 \pm 0.07$. Adding two PMS binaries recently analyzed (P2494 and P2445 in Orion, unpublished CORAVEL measurements, see Table 2 in Sec.4), we get $\bar{e} = 0.31 \pm 0.05$.

iii) the mass ratio distribution ($q = M_2/M_1$), including all kinds of binaries in the unbiased sample (SB, VB, CPM), is a regular function increasing toward low-q values and down to $q = 0.25$, probably decreasing below that value. Although [DuM91] found a rather flat distribution for SBs with $P < 1000d$, they though premature to claim a difference with the distribution for longer periods. However Mazeh and Goldberg ([MaG92], this volume) presented a new and elegant method to analyse the mass ratio distribution from the mass function of the SBs. Applying this method to the sample of nearby G-dwarfs, Mazeh et al. [MGD92] find that the mass ratio distribution for SBs with $P < 3000d$ is a slightly rising function toward $q = 1$, although a flat distribution remains possible due to the small number of binaries involved.

2.2. Among open clusters

2.2.1. Sample and observations

The dwarfs with spectral types later than F5 in the nearby open clusters (Hyades, Pleiades, Praesepe, Coma Berenices and α Persei) define samples with well defined ages and chemical composition. It will be interesting to compare the properties of the cluster stars to those of the field nearby and halo star samples. In 1978, we started an observing program including all known main-sequence members later than F5 in the four large northern open clusters α Persei, Pleiades, Praesepe, and Coma Berenices. Later, NGC 752 was added to this sample. Systematic observations of the Hyades cluster have been obtained for many years by Griffin and coworkers [GGZ85], and by Latham and collaborators [StL85], so we did not make extensive observations of this cluster. The main body of the observations is completed and most orbits have been determined. Publication of the results has started with the discussion of the coronas in Praesepe [MWD90], and the Pleiades [RMM92]. A paper on the analysis of 11 spectroscopic binaries in the Pleiades has been submitted [MRD92] and will be followed by similar studies for the other clusters.

The limiting magnitude of the 1-m Swiss telescope installed in the Haute-Provence Observatory (France) is around $B = 12.5$ for "routine" observations. This limit corresponds to stars of spectral type around K0 in these clusters which are at similar distances. The observing conditions and reductions have been described in other papers [MWD90], [RMM92] and are similar to those of the nearby G-dwarfs. Stars fainter than this limit have been observed, but the completeness of the fainter sample cannot be guaranteed since the selection was made according to specific purposes other than completeness. In fact, the programme was not restricted to the observations of known members, we tried to find as many new members as possible. In 1983, a similar programme was started with the CORAVEL scanner installed on the 1.54m Danish telescope at the ESO Observatory located at La Silla (Chile) which includes the southern nearby clusters: IC 2391, IC 2602, NGC 6475 and Blanco 1. The limiting magnitude of the 1.54-m Danish telescope is around $B = 14.5$.

The total number of stars thus included in the main-sequence observing programme is presently close to 900. 75 binaries have been discovered and observed continuously until a satisfactory orbit could be computed. So far, orbits for 46 members have been determined. However, some binaries turned out to be non-members after the systemic velocity was computed, especially in the α Persei and Coma Berenices clusters, where the membership determination based on proper motion is less selective. Table 1 gives the number of stars (column 2) observed in each cluster listed in column 1, as well as the number of members (3), triple systems (4), double-lined spectroscopic binaries (5), single-lined spectrocopic binaries (6), photometrically analyzed binaries not detected by their radial velocities (7), visual binaries if any (8), and the number of orbits already obtained (9). Column 10 gives the number of stars in the magnitude range in which the sample is complete. The next column gives the percentage of spectroscopic binaries. The information on the clusters in the lower part of the table is less elaborated than that for the other clusters. The clusters are discussed separately in the following sections.

i) Praesepe

The main survey undertaken in Praesepe contains 90 known members from Klein-Wassink's study [KW27]. We have discovered 23 binaries and three triple systems which present quite different characteristics. KW 495 is the most obvious triple system since three correlation dips are clearly visible with CORAVEL over long timescales. The central dip is at a constant velocity (the cluster velocity), while

Table 1. Number of observed stars (2) and detected binaries (5,6) in the nearby open clusters. For the signification of other columns, see text.

(1)	(2)	(3)	(4)	(5)	(6)	(7)	(8)	(9)	(10)	(11)
Pleiades	100	100	2	4	8	13	4	11	85	13%
Halo I	83	56			3	5				
Halo II	101	30								
Praesepe	90	85	3	4	16	10	1	17	80	29%
Halo	117	48		2	8	3		8		
Coma Ber	64	24	1	3	3	2		7	22	23%
α Per	47	24		1	2	4		2		
IC 2391	11	11		2	3					
IC 2602	11	9								
NGC 6475	33	27		1	1					
Blanco 1	116	54								
NGC 752	86			4	9			6		
NGC 1976	38			1						

the other two, corresponding to a double-lined binary, are disposed symmetrically around the constant one. KW 365 is a binary with a rather eccentric orbit and shows a double correlation dip at the phases of maximum elongation: one dip at a constant velocity appears superposed onto the variable dip. Finally, KW 367, a short period double-lined binary, shows a long term variation of its systemic velocity. All three systems are highly hierarchical. In the first case, all three stars are visible. In the second one, the single star and the primary are visible and in the third, both components of the binary are visible, but not the "single" star.

In addition to the observation of Praesepe's central part, we have looked for new members in the cluster corona and observed 117 stars selected from the proper motion studies of Artjukhina [Art66a], [Art66b]. The area investigated by Artjukhina has a radius of 4 degrees which is larger than that of any other proper motion study ever made in this cluster. We confirmed the membership of 48 stars [MWD90] and discovered 10 spectroscopic binaries. Six orbits were determined, with periods from 1 to 143 days. One more star should be added to the statistics presented in Table 1, namely KW 244, a short period eclipsing binary [Wil76], [LeH83] which did not produce any correlation dip, probably due to the large rotation induced in the stars by the short orbital period.

ii) Pleiades

The main body of the initial survey contains 92 stars from Hertzsprung's study [Her47]. Thirteen spectroscopic binaries were discovered and eleven orbits determined, with periods smaller than 1000d. One triple system has been discovered: HII 2027. This system is similar to KW 365 discussed above: it shows one dip at a nearly constant velocity, corresponding to a single star (A), and a second one moving with a period of 48.6d, corresponding to a single-lined binary (B). After twelve years of observations, the systemic motion around the center of mass of A-B (i.e. the long period motion) has been observed [MRD92].

We have also observed 83 stars in the corona of the Pleiades selected from van Leeuwen's et al list [vLA86], which is based on unpublished proper motions by Pels, and Walraven photometry. Our analysis of the radial velocities confirmed the membership of 56 stars only [RMM92]. Three binaries were discovered in this sample, which results in a unusually low frequency (5%). However this sample does not represent a complete census of corona members down to the limiting magnitude of the observing list. To improve the knowledge of the cluster stellar content and structure, 100 new candidates selected from the list of Artjukhina and Kalinina [AKa70] in the Pleiades' corona have been observed in December 1991, and some 30 additional candidate members have already been revealed.

iii) Coma Berenices

In the Coma Berenices cluster, all 19 late type members from Trumpler's [Tru38] extensive study were observed. 45 stars were added to the survey according to different selection criteria. First, we tested a number of stars for which Trumpler did not have radial velocities, and stars with slightly discrepant data which had not received the highest rating and therefore were not considered to be members. This allowed for the confirmation of the membership of two additional stars, Tr 12 and 36. A number of stars were proposed by Olsen [Ols84] for cluster membership, some were selected from the photographic study of Argue and Kenworthy [ArK69] and a few candidates were retained from the study of DeLuca and Weis [LuW81].

Finally, in collaboration with M. Grenon (Geneva Observatory), 22 additional stars were selected, on the basis of their magnitudes and colours, from the surveys made by Malmquist [Mal27], [Mal27] in the region of the galactic pole. The total number of stars observed finally amounts to 64. As was the case in Praesepe, one short period binary (Tr 111, P = 0.96d) could not be observed with the CORAVEL scanner due to its large rotation. The orbit has been computed by Kraft [Kra65].

iv) α Persei

The α Persei sample was based on the stars which were confirmed members by the UBV photometric observations [Mit60]. It was enlarged with a number of additional stars taken from several sources. The second selection is based on proper motions [Fre80], uvby photometry [TRJ89] and proper motions and radial velocities [SHB85]. The total number of stars observed amounts to 47, but the completeness of this observing list is difficult to estimate. Few binaries have been discovered and the membership of two of them is questionable since the systemic velocities differ by some 4 to 5 km/s from the mean cluster velocity. Only one double-lined binary (No 848) has been found among the cluster members and its orbit has been obtained. Several non-members have been detected, even among stars usually considered as cluster members on the basis of their photometry and proper motions.

v) NGC 6475

The sample observed in NGC 6475 contains 13 dwarfs later than F5 selected from Koelbloed's list [Koe59], and 20 stars selected on the basis of proper motions and photographic UBV photometry [CHN69]. We obtained successful results since 27 stars among the 33 observed have precisely the expected velocity for the membership to NGC 6475. Two observations made one year apart allowed the discovery of only one spectroscopic binary. The main sequence of the cluster, which ended at $V = 11.5$ according to the photographic observation of Hoag et al [HJI61], has now been extended to almost $V = 13$.

vi) Blanco 1 (Zeta Scl)

The main-sequence extension of this nearby cluster (200 pc) first studied by Epstein [Eps68] and Eggen [Egg70] was also rather short and stopped at about V = 10.5. An observing program has been undertaken to extend the list of members towards fainter magnitudes. For the most part, the candidates have been selected from the studies of Abraham de Epstein and Epstein [AEE85] and Westerlund et al. [WGL88]. The total sample contains 117 stars and the membership of 54 stars has been confirmed. The main sequence is now well defined and extends to $V = 14.0$. The corresponding colour index is $(B - V) = 1.0$. Three to four observations have been secured for each star, but few binaries were found. More observations are needed to complete the detection. Due to the position of the cluster (in the direction of the south galactic pole) and its small distance, the surface area occupied by the cluster over the sky is rather large and it is very probable that the area which have been investigated photometrically is not large enough to cover the whole possible (tidal) radius. Accordingly, it is difficult to estimate the degree of completeness of the present sample.

vii) NGC 752

We observed 86 stars in the open cluster NGC 752 with the CORAVEL scanner. Initially, the main motivation was to prove that the cluster main-sequence does not end at about $V = 12$ as resulted from the study of Johnson [Joh53].
The selection was based on the UBV photoelectric and photographic photometry available in the database for stars in open clusters [Mer88] and [Mer92], for stars in Heinemann's list [Hei26]. Our first selection took into account the proper motions obtained by Lavdovski [Lav61]. The membership has been revised according to recent proper motion studies ([Fra89], [Pla91]). The selection of candidates fainter than $V = 12$ is based on the photographic UBV study of Rohlfs and Vanysek [RoV61] and the proper motion membership probabilities published by Francic [Fra89] and

Platais [Pla91]. Our radial velocity observations clearly show that the main sequence extends to magnitudes fainter than $V = 12$ and is populated normally. Several spectroscopic binaries have been independently discovered by R.D. Mathieu and the authors. A number of orbits have already been determined and both data sets are complementary in the sense that the orbits obtained by each group are for different stars. This programme is being continued in collaboration with R. D. Mathieu and the combined sample will contain more than 110 stars.

viii) IC 2391 and IC 2602

These two similar, poorly populated, southern clusters were put on the observing list due to their short distances from the Sun (around 150 pc for each). They present potential interest to study main-sequence stars in clusters younger than the Pleiades. However, despite their proximity, few late-type members are presently known. The proper motion and photometric study reported by Stauffer et al [SHJ89] added only a couple of new members. So far, 11 stars were observed in each cluster with the southern CORAVEL scanner. Five spectroscopic binaries have been discovered in IC 2391 and none in IC 2602.

ix) Orion cluster

Following the publication of rotational velocities for a sample of Orion population F-G dwarfs by Smith et al [SBB83], which shows an appreciable number of slow rotators, we undertook an observing program to obtain radial velocities for slowly rotating stars in this sample and thus check their membership. A large fraction of the slowly rotating stars have been rejected for membership on the basis of the radial velocities. Preliminary results have been published by Mermilliod and Mayor [MeM85].

2.2.2.   The binary rate

At first sight, the percentage of binaries with periods shorter than 1000d in the Pleiades ($11/85 = 13\% \pm 4\%$) is slightly less than that of Praesepe ($16/80 = 20\% \pm 5\%$). However, the total fraction of photometrically analysed binaries detected with the condition that the distance above the ZAMS is larger than 0.10 mag is very similar, $25/85 = 29\% \pm 6\%$ in the Pleiades and $26/80 = 33\% \pm 6\%$ in Praesepe. These statiscctics have been made in the colour region where the sample is considered to be complete, i.e. $0.40 < (B - V)_0 < 0.85$. A detailed study of the Pleiades made by Mermilliod et al [MRD92] showed that there is no unique relation between the binaries detected spectroscopically and those detected photometrically. Radial velocity observations failed to detect 70% of the photometrically analysed binaries (a marginal detection at the 5% level is possible in a few case), even though the stars present an evident effect in the $V, (B-V)$ diagram due to the presence of a companion. Conversely, 50% of the spectroscopic binaries are not seen photometrically, which implies that the secondary star is fainter by about 5 mag (the resulting effect on the

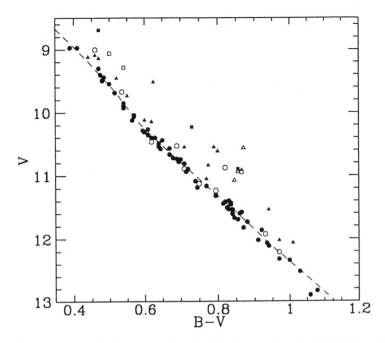

Figure 1. Color-magnitude diagram for the F-G stars in the Pleiades. The symbols identify the single stars (filled circles), SB1 (open circles), SB2 (filled squares), visual binaries (open squares), triple systems (open triangles) and photometrically analyzed binaries (filled triangles).

V-magnitude and $(B - V)$ colour is less than 0.01 mag). Figure 1 shows the F5-K0 main sequence of the Pleiades with various symbols indicating the duplicity status of each star. Further observations with other techniques would be very useful to uncover the nature of the secondary stars.

2.2.3.   The e vs $\log P$ diagrams

The orbital elements of the spectroscopic binaries in α Persei, the Pleiades, Praesepe and Coma Berenices clusters derived from CORAVEL observations and those published for the Hyades [GrG78], [GrG81], [GMG82], [GGZ85] are used to discuss, with more data than in 1984, the $(e, \log P)$ diagrams at two different ages. We have grouped the data for the Pleiades and α Persei clusters on one side (Fig.2a) and those for the Hyades, Praesepe and Coma Berenices clusters on the other side (Fig.2b). The emerging pattern is similar for both age. Orbits with periods shorter than about $\log P = 1$ are mostly circular, while the longer period orbits have non zero eccentricities. The limit is not well constrained for the Pleiades due to the lack of stars with periods close to the limiting period. Conversely, the cut-off is well marked

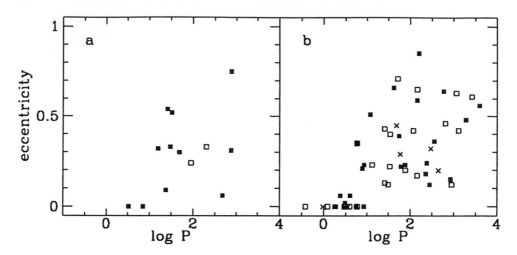

Figure 2.   Comparison of the $(e, \log P)$ diagrams for two cluster generations: (a) binaries in the Pleiades (filled squares) and in α Persei (open squares), (b) binaries in the Hyades (filled squares), Praesepe (open squares) and Coma Berenices (crosses).

in the Hyades-Praesepe-Coma diagram. The latter diagram shows a number of stars with orbital periods shorter than the cut-off period, but still eccentric orbits. These exceptions are due to incomplete tidal circularization as is explained in section 3.4 of this paper. It is remarkable that there exists a lack of systems with orbital periods in the interval $1 < \log P < 3$ with eccentricity smaller than 0.1. The same feature has been noticed for the open cluster red giants (MeM92, this volume).

2.2.4.   The eccentricity distribution

The available data for the F-G binaries in these five open clusters, with periods larger than the cut-off period and less than 1000d (43 stars), have been grouped together to obtain the distribution of eccentricity (Fig.3). The lower limit has been chosen to remove from the sample the stars with circular orbits (18) since they no longer reflect the original conditions, and the upper one corresponds to the longest period for which we hope to have an unbiased sample. The decrease of the number of orbits with large eccentricities may nevertheless be partly due to observational biases, the difficulty of detecting radial velocity variations for binaries with long periods, and high eccentricity.

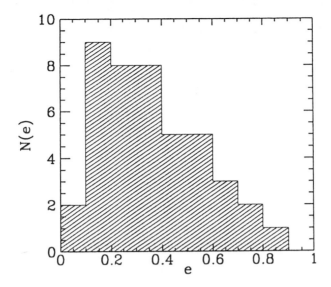

Figure 3. Eccentricity distribution for F-K dwarfs in open clusters with $10 < P <$ 1000d

2.2.5.   The orbital period distribution

The whole sample (75 stars) has been used to derive the distribution of periods (Fig.4). Of course the sample is not complete, but the additional data should not change the distribution too much. Most missing orbits have long periods and will populate the right side of the distribution. In the Pleiades, we have two orbits to solve: one is long ($P > 6000$d) and the other is not yet defined due to the small number of observations obtained so far. However, the amplitude seems to be small. In Praesepe, there are still 8 stars without an orbit determination. Four have quite certainly an orbital period shorter than 1000d and the other four have a longer one. The orbital parameters for the remaining binaries in the Hyades will, however, greatly improve the distribution.

3.   Evolution of the orbital elements

3.1.   Some evolution processes

The orbital elements of SBs with $P < 3000$d can be altered at various stages of their evolution. Alteration processes include:

i) tidal effects. They are found to be specially effective among stars with convective envelopes due to the short timescale of their turbulent convection relatively to, and interacting with, the tidal distorsion timescale ([Zah77], [Zah89]), itself related to the

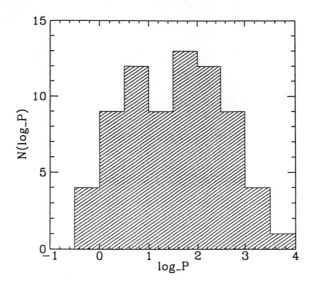

Figure 4. Distribution of the orbital period for 75 F-K dwarfs in the nearby open clusters.

orbital period. These effects tend first to synchronize, then to circularize the orbits. They affect primarily SBs with $P \leq 10d$ on the PMS and on the MS, or $P \leq 100d$ and $P \leq 1000d$ for giant and supergiant SBs respectively.

ii) angular momentum loss by magnetically coupled stellar wind in orbit-spin coupled binaries (AMLOSC) binaries, a mechanism proposed by several authors (e.g. [Egg86], [VeM92]). The magnetic activity generated in the convective zone of late-type stars lead them to loose angular momentum through solar winds trapped in the corotating magnetic field. When such a process is met in synchronized binaries, this angular momentum loss is made at the expenses of the orbital momentum reservoir. This causes the orbital period to decrease and the star to spin up. According to [VeM88], this process affects the SBs with $P \leq 4d$, and the continuing process probably leads to the coalescence of the components.

iii) angular momentum loss by gravitational wave radiation (GWR). Such radiation is possible in the presence of large masses at small distances, so it affects mainly very short period systems with degenerate components. The timescale of the orbital period decrease due to GWR may be as short as to be actually detected in a pulsar like PSR 1916+34 (see e.g. the review by Tutukov and Yungelson [TuY87]).

iv) strong wind accretion. A star evolving through the asymptotic giant branch will expulse its envelope as a strong wind. A relatively far companion (far enough to survive the evolution of the primary, say $P \simeq 10^2 - 10^4 d$) is in position to accrete this wind. This is the scenario explaining the overabundances in s-elements observed in the atmosphere of many (if not all) Ba II and CH stars, observed now as giant primaries accompanied by a relatively distant and probably white dwarf secondary responsible of the above process (Boffin and Jorissen [BoJ88], Jorissen and Boffin [JoB92]). As a search for the progenitors of these Ba giants, see also North and Duquennoy [NoD92].

v) Roche-lobe overflow and mass transfer, also occuring when one or both of the components in a close binary system leave the MS. Clearly the distributions of orbital elements are strongly affected in this process. Various evolutionary path for moderately wide binaries are proposed by Eggleton [Egg86], [Egg92].

3.2. The particular case of tidal effects

In the following, we will restrict to a simple description of the evolution of the elements (P, e) in the context of tidal effects. This work is justified by the present status of the confrontation between tidal theories and observations. It is intended to explore the apparent inconsistencies within short-P eccentric orbits, as well as the general behaviour of the circularization process on the MS which could be extrapolated on the PMS.

We have written a computer programme simulating the evolution of the orbital elements (P, e) of a late-type dwarf binary under tidal effects. We choosed the framework of Zahn's (1977) theory involving tidal dissipation in MS stars with convective envelopes. Taking equations 4.3 and 4.4 of [Zah77] giving the evolution of the semi-major axis da/dt and of the eccentricity de/dt, we assume:

i) synchronism of the rotational and orbital velocities ($\Omega = \omega$) This is probably the case of most of the observed MS binaries affected by tidal stresses, since Zahn estimates that the synchronization time for a 10d-period SB is about 1000 times shorter than the circularization time.

ii) small initial eccentricities ($e < 0.3$). This allows to neglect 4th order terms in e and to make valid the already quoted Eq. 4.3 and 4.4.

iii) clock hypothesis. The circularization timescale t_{circ} of a MS binary of period P can be expressed in the form $t_{circ} = AP^{16/3}$ (from Eq. 6.2 of [Zah77]). For coeval binaries, their age can be assimilated to the time needed to circularize the longest orbit observed now as circular. For solar-mass stars on the MS, A is roughly a constant which can be determined by the study of open clusters. In this paper we calibrate A with the cluster M67, for which the tidal effects clearly circularize the orbits up to a reasonably well defined period around 10.5d [MLG90]. The age of M67 is taken to 5 Gyr from fitting isochrones (Maeder and Mermilliod [MaM81]). The choice of M67 to calibrate A is somewhat arbitrary. It depends on the credibility given to P_{cut} as representative of the true cut-off period (see a discussion by [MDL92]). Let us recall that the parameter A also contains a dependence in R^{-8} where R is the stellar radius, which is significantly larger in the PMS stage and should decrease considerably the circularization timescale on the PMS [MaM84], [Zah89]. We also stress that the chosen exponent $g = 16/3$ corresponds to a minimal action of the tidal effects on the orbital elements, compared to exponents $g = 13/3$ (as it can be derived from [Zah89]) or $g = 10/3$ (as in [GoM91]).

Using these asumptions and Kepler's law relating a to P, we get the following set of simplified equations:

$$t_{circ} = AP^{16/3} \qquad (1)$$

$$\frac{de}{dt} = \frac{-e}{t_{circ}} \qquad (2)$$

$$\frac{dP}{dt} = \frac{-57}{7}\frac{Pe^2}{t_{circ}} \qquad (3)$$

Some interesting results from the simulated tidal effects can be derived:

i) Eqs 2 and 3 lead to the following relation valid at each t:

$$ln\frac{P(t)}{P_0} = \frac{57}{14}(e^2(t) - e_0^2(t)) \qquad (4)$$

which implies that the orbital period varies drastically when the eccentricity is high. We plot in Fig.5 this function for different values of e_0. The simulations being essentially valid for small eccentricities, the conditions $e_0 = 0.54$ and $e_0 = 0.90$ are just plotted for qualitative interpretations. With this in mind, it appears that the initial orbital period P_0 can be greatly reduced throughout the circularization process, by a factor of 1.45 to 25 for $e_0 = 0.3$ to $e_0 = 0.9$ respectively, before reaching $e = 0$. We stress that Eq.4 is independent of the value of the exponent $g = 16/3$ used in Eq.1. It means that if g is modified (e.g. $g = 10/3$ in [GoM91]), the shape of the variation curves in Fig.5 remains qualitatively the same, and only the timing of the variations is changed.

ii) Fig.5 is not labeled in time, however the circularization timescale is strongly dependent on the period P_0. Integrating Eqs 1.a to 1.c, with a variable time step adapted to the variation rate of the eccentricity, we get the time-labeled evolutions for various initial conditions (P_0, e_0). In particular, we derive that a binary arriving on the MS with $(P_0, e_0) = (11.6d, 0.54)$ reaches a point $(5.86d, 0.35)$ after about 0.8 Gyr. The latter coordinates are precisely the observed values for the binary KW 181 in Praesepe at the same age. A representation of its evolutionary path is given in Fig.8. This binary was formerly taken as representative of the tidal cut-off period [MaM84], [MLG90]. We see here that its eccentricity can be explained in terms of incomplete tidal circularization, which should last for KW181 still about 0.1 Gyr before being completed. Consequently, this kind of still eccentric binary is not in contradiction with the theoretical view of Zahn and Bouchet (1989) that tidal effects predominate

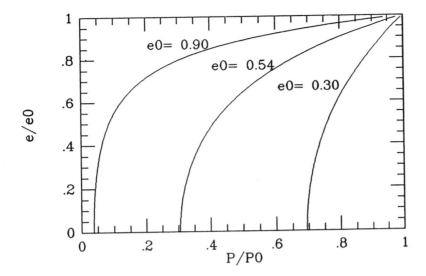

Figure 5. Evolution of the orbital elements e, P relatively to their initial values, according to the MS tidal effects simulated in this paper (Eq.4). The results for initial conditions $e_0 > 0.3$ are only qualitative due to the approximation adopted in our simulations (see text).

on the PMS and are able to circularize essentially all binaries born (at $t = 0$) with P up to 7-8d.

iii) The above result leads to reconsider the definition of the tidal cut-off period distinguishing the circular from the eccentric orbits, as formerly being the period of the shortest period eccentric orbit. We propose here to redefine this cut-off as the period of the longest period circular orbit. Due to the discrete nature of the binary samples this definition represents a lower limit to the true cut-off period, but it seems justified by the following facts:
— it avoids to misinterpret any shorter period binary, still eccentric due to incomplete tidal circularization (this paper), or to dynamical interactions with a third body present in the system which can prevent the total circularization of the inner binary (mechanism studied by Mazeh and Shaham [MaS79], invoked by Mayor and Mermilliod [MaM84], and reinvestigated by Mazeh [Maz90]).
— the definition assumes homogeneous samples, in particular of stars with same mass and evolutionary status in the HR diagramme. Circular orbits with much longer periods than 10 days are generally associated with evolved primaries or degenerate secondaries.
— it seems improbable that this cut-off period can be confused with an original, circular-born binary, in view of the apparent absence of long-P (> 20d) circular orbits among the samples of unperturbed binaries observed to date.

iv) We can apply this evolutionary scenario to all the binaries in a coeval sample, to see how they appear in a $e/\log P$ diagram at various stages of their tidal evolution on the MS. But before doing this in Sec. 3.4, let us include another process of orbital element alteration affecting short-P binaries, the AMLOSC.

3.3.   The case of AMLOSC

According to the scenario of Van't Veer and Maceroni, see e.g. [VeM88], the AMLOSC process in SBs leads to a shrinking of the orbit and to a corresponding decrease in orbital period.   Various period evolution fonctions (PEFs) are proposed by these authors. We choose in our study the PEF given by Fig.4 of Maceroni and Van't Veer [MaV91], which we tabulated to include in the above simulation programme. The effect is included when the period decreased by tidal effects reaches the value of about 3.46d ($\log P = 0.54$). Below $P \simeq 0.2$d ($\log P = -0.7$), we assume rapid disparition of the binary under its classical way of detection. Although this process has been given less development in this paper than tidal effects, we invoke it to explore the possibility of explaining the deficiency in short-P binaries among old samples such as the galactic halo (Latham et al. [LMC88], Torres et al. [TLM92]).

3.4.   An evolution of orbital elements for solar-mass MS stars

Fig.6 shows the simulated evolution with combined tidal effects and AMLOSC, of a sample of coeval binaries taken here as all the solar mass MS binaries contained in the three clusters Hyades, Praesepe and Coma Ber assumed to have roughly the same age (say 1 Gyr). For this sample of 53 SBs with $P < 1000$d at $t = 1$Gyr, the observed tidal cut-off period is $P_{cut} = 8.5$d, and the number of binaries with $P < 10$d is $N_{10} = 23$, or a rate of 43%.

At $t = 5$Gyr, we still have $P_{cut} = 8.5$d but 5 binaries with $e_0 \simeq 0.3$ are now circularized and 10 binaries have "disappeared". We have thus $N_{10} \simeq 13$ or a rate of $13/43 = 30\%$ of short-P binaries, significantly closer to the $3/13 = 23\%$ observed in M67 with the same age and studied by [MLG90].

At $t = 15$Gyr, P_{cut} is on the way to skip from 8.5d to about 10.6d with the star No.127 in Praesepe being nearly circularized. But formally, P_{cut} remains unchanged which illustrates the problem of small samples of binaries. Moreover, there is a gap between this 10.6d-period binary and the next one at 25.8d which will poorly constrain the true cut-off period. For the halo, the true cut-off period is situated somewhere between 12d and 19d [LMC88], [JaM88]. Finally, the rate of short-P binaries is $N_{10} = 10/43 = 23\%$ to be compared with the very low rate of $3/40 = 8\%$ in the halo inferred from [TLM92].

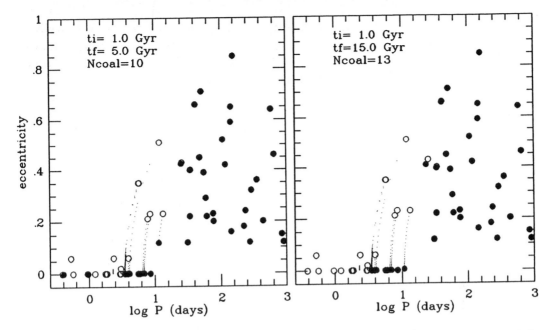

Figure 6. Evolution of the orbital elements in the plane $(e, \log P)$ according to tidal effects and AMLOSC processes simulated in this paper, for the Hyades-Praesepe-Coma sample. The simulation shows the distribution obtained at various epochs of its evolution, between t_i (open circles) and t_f (filled circles)

In conclusion, there is a qualitatively good agreement between the scenario proposed and the observations of various coeval sample: potential increase of P_{cut} with age (i.e. persistence of detectable tidal effects on the MS) and AMLOSC. However we note the difficulty to deal with small binary samples: this will be partly overcome in Sec.4 by merging all the binaries of known age and all kind of eccentricities in a same diagram. In addition, the AMLOSC scenario only partly explains the deficiency of short-P binaries in old samples.

4.   Discussion

One important diagram in the field of tidal circularization in close binaries is the $e/\log P$ diagram (see previous section and several contributions in this volume). Another one is the $\log P/\log t$ diagram where P is again the orbital period of the binary and t its age.

Until now, this later diagram contained only the few cut-off periods (P_{cut}) observed in several samples of known age, where this cut-off was defined as the shortest period eccentric orbit. Different studies ([MaM88], [GoM91]) have tried to conciliate these sparse data with the theory . The latest proposed a model where the turbulent viscosity, responsible for the dissipation of the tidal effects, depends quadratically on the orbital period, leading to a dependence of the circularization timescale on $P^{10/3}$. According to [GoM91], this exponent $g = 10/3$ would fit better the observations, with the above definition of P_{cut}, than the exponent 16/3 of Zahn [Zah77].

However, we have seen in Sec.3 how not only e but also P varies with time under the action of tidal effects and how this can lead to a misinterpretation of P_{cut}. This was illustrated by the case of the binary KW 181 in Praesepe. Consequently the above data on various P_{cut}, being already sparse, are moreover probably misinterpreted.

Displaying only P_{cut} in the $\log P / \log t$ diagram looses the information on the great number of accumulated binaries, in particular on their eccentricity. In this paper we display all the solar-mass binaries of known age available to date (including the most recent analyses presented in this volume), divided into two populations according to their eccentricity (Fig.7). The sample names, sources of orbital elements and ages are given in Table 2. The ages of open clusters have been reevaluated by JCM by fitting new isochrones from Schaller et al. [SSM92]. The dividing eccentricity is chosen here to be $e = 0.03$. According to Mazeh [Maz90], the presence of a third component can induce a substantial eccentricity in the inner binary system due to dynamical interactions different from tidal effects. So in Fig.7, we removed the short-P binaries belonging to known triple systems, which may pollute the diagram. Another particular case is the galactic halo binary G171-23, not appearing on Fig.7 due to the adopted vertical scale, but circularized with a period of 153d [LMC88]. Its mass function $f(m) = 0.0815 M_\odot$ is consistent with the presence of a degenerate companion of mass $0.5 - 0.6 M_\odot$, if we assume a primary mass of about $0.8 M_\odot$ suggested by the colors given by [LMC88]. Thus we suggest that the companion of G171-23 is indeed a white dwarf, that the orbit was circularized when the present secondary was at the giant stage where stellar radii are much larger and tidal effects affect longer periods, and that this orbital period is not related to the true cut-off period for MS stars.

The figure now obtained supports the need to adopt the new definition of P_{cut} as the longest period circular orbit in a homogeneous coeval sample. In addition, Fig.7 still suggests, admitting the new definition of P_{cut} in Sec. 3.2, the existence of a variation (increase) of P_{cut} with the sample's age, which consequently seems to validate a tidal scenario similar to that of [Zah77]. However, we show below that the presently available observations are not inconsistent with the scenario proposed by Zahn and Bouchet [ZaB89], in which the circularization processes for late-type dwarfs have little effect on the elements (P, e) on the MS, while they predominate during the PMS stage and circularize the orbits up to original $(t = 0)$ orbital periods of about 8d.

On the same diagram, we have plotted straight lines corrresponding to two proposed theoretical variation laws of P_{cut}: the one in $P^{16/3}$ (continuous line, [Zah77]), and the one in $P^{10/3}$ (dashed line, [GoM91]), both scaled to M67 cut-off period as defined in Sec.3. As by definition of P_{cut} both lines must run above each observed circular orbits

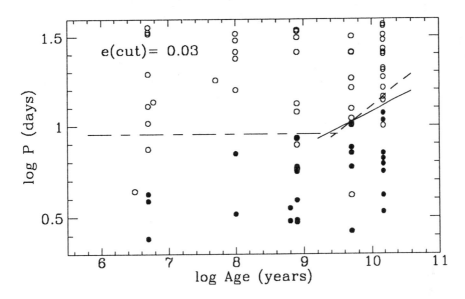

Figure 7. Diagram $\log P$ vs $\log Age$ for solar-mass spectroscopic binaries of known age. Symbols distinguish the orbits considered circular (filled circles: $e < 0.03$) from those considered eccentric (circles: $e > 0.03$). Straight lines correspond to the location of the circularization period P_{cut} in different tidal theories: continuous: $g = 16/3$ [Zah77], dashed: $g = 10/3$ [GoM91], short dash-long dash: PMS circularization up to $P \simeq 9$d (see text).

Table 2. Sources of orbital elements for the samples of non evolved solar mass binaries plotted in Fig.7

Sample	Sources	Age (yr)
PMS	Mathieu [Mat92] and references therein	$3 \times 10^6 - 10^7$
	Unpublished CORAVEL measurements for P2494	
	($P = 19.5$d, $e = 0.26$) and P2445 ($P = 119$d, $e = 0.23$)	
α Per	Mermilliod and Mayor (in preparation)	5×10^7
Pleiades	Mermilliod et al. [MMM92]	1×10^8
Hyades	Works of Griffin and Gunn (e.g. [GGZ85])	8×10^8
Praesepe	Mermilliod and Mayor (in preparation)	8×10^8
Coma Ber	Mermilliod and Mayor (in preparation)	6×10^8
M67	Latham et al. [LMM92] and references therein	5×10^9
Halo	Latham et al. [LTS92] and references therein	$10 - 15 \times 10^9$

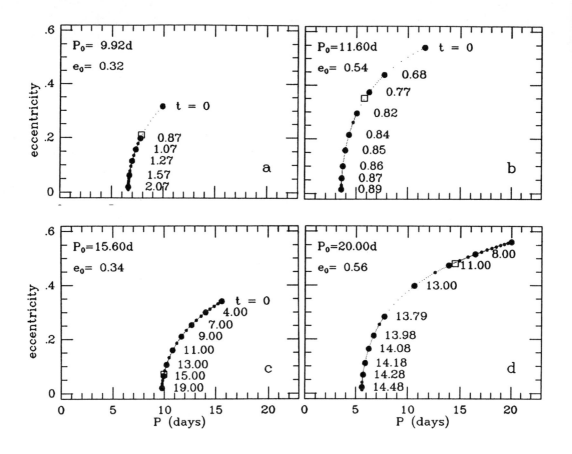

Figure 8. Evolution of the orbital elements in the plane e vs P according to tidal effects simulated in this paper for some of the binaries in Table 3. The curves are labeled with time in Gyr for large filled circles. The open square indicate the observed position. The model shows that eccentric MS binaries can be found with periods below the true cut-off period associated with each coeval sample considered in the text, due to incomplete tidal circularization.

Table 3. Evolutionary timescales of orbital elements (P, e) according to tidal effects simulated in this paper, for incompletely circularized binaries. An exponent $g = 16/3$ was adopted, except * : $g = 13/3$, ** : $g = 10/3$. Sources for orbital elements are: (1) [Mat92], (2) [GGZ85], (3) [GrG78], (4) Mermilliod and Mayor (in preparation), (5) [Mil91], (6) [LMC88]

Sample	Star name	Observed		Initial		Time (Gyr) to reach:	
		P (d)	e	P_0 (d)	e_0	observed e	$e = 0.01$
PMS	EK Cep [1]	4.43	0.11	9.00	0.431	0.403	0.502
	EK Cep *			9.00	0.431	0.059	0.082
	EK Cep **			9.00	0.431	0.009	0.014
Hyades	vB 164 [2]	7.86	0.21	9.92	0.316	0.80	2.37
	vB 121 [3]	5.75	0.35	11.60	0.54	0.80	0.89
Praesepe	KW 181 [4]	5.86	0.35	11.63	0.54	0.81	0.91
M67	S 1284 [5]	4.18	0.27	17.7	0.65	5.02	5.06
Halo	BD +05°3080 [6]	9.94	0.07	15.6	0.34	14.65	> 20
	G87-47 [6]	13.73	0.46	20.0	0.56	11.2	14.6
	HD 118981 [6]	14.50	0.48	20.0	0.56	10.5	14.6
	HD 115968 [6]	16.20	0.28	18.5	0.33	12.8	> 20

(filled circles in Fig.7), we represented a third straight line at a constant $P_{cut} \simeq 9d$ to allow for uniform PMS circularization. This value is not far from the theoretical prediction of [ZaB89], however the present data allow even slightly higher values, up to $\simeq 10d$ as will be shown below. We may indeed ask the question whether every still eccentric orbits with $P < P_{cut}$ can be explained with incomplete tidal circularization as for KW 181 in Sec.3. The computer programme permits to play back in time the evolution of (P, e) for each observed eccentric binary on the MS, assuming the use of $g=16/3$ represents a reasonable approximation of the circularization timescale on the MS. We find that the derivation of the initial elements (P_0, e_0), with the conditions $P_0 > P_{cut}$ and a running time equal or less than the observed age, is straightforward for all such binaries in Fig.7. These elements are displayed with the corresponding evolution times in Table 3, and Fig.8 shows some examples of the simulated evolutionary paths.

One exception is EK Cep which is not yet on the MS. Its age estimate ranges from 3.10^6 to 2.10^7yrs [MDL92]. We do not know P_{cut} for this age, but in the context of Zahn and Bouchet's theory it is expected to be above 7d. With the equations 1 to 3 of Sec.3 and the calibration of A as in Sec. 3.2, and to evolve from an initial period around 8d in a time less than its age to the present $(P, e) = (4.4d, 0.11)$, EK Cep has to come from e.g. (9.0d, 0.43) in 0.4 Gyr which is too long. However, according to Zahn [Zah77], [Zah89], the circularization timescale varies with the stellar radius as the inverse of its 8th power. So the above evolution time can be greatly reduced, e.g. by 256 if the PMS radius of EK Cep is twice that it would have on the MS, for which our simulations were calibrated. We also tested the exponents $g = 13/3$ and $g = 10/3$ in our programme, instead of $g = 16/3$. They gave naturally much shorter timescales (see Table 3), but in view of the above remark on PMS stellar radii, it is not necessary to invoke them to explain the present status of EK Cep. Moreover, the simulations made by Lubow and Artymowicz [LuA92] indicate that PMS disc-binary gravitational interactions may induce substantial eccentricity, which thus can counteract the tidal effects.

In conclusion, even with the high value of $g=16/3$ which is presently the less efficient way to circularize binaries, EK Cep does not bring any constraint on the tidal theories. More generally, the available data on binaries of age less than 0.8 Gyr presently seem to bring no constraint on the values of the coefficients A and g or on P_{cut} in this range of ages. Consequently they are not inconsistent with the suggestion that $P_{cut,PMS}$ can be as high as $\simeq$ 9d. One strong constraint would be to find a circular PMS binary with a period around 6-9d. More generally, any young binary ($t < 0.1$Gyr) with $P \simeq$ 7d may give important indications on the tidal processes.

5.   Conclusions

We have reviewed several radial velocity surveys currently under acquisition with the CORAVEL spectrometers in the field of stellar duplicity, among solar-mass stars in the field and in open clusters. We attempted to use the binary information contained in these surveys to test the current theories of stellar tidal circularization, in particular using the evolution equations of orbital elements (P, e) in [Zah77]'s theory.

We have shown that eccentric orbits with period shorter than the former definition of the tidal cut-off period can be explained by incomplete tidal circularization. This lead us to propose a new definition of this cut-off as the period of the longest period circular orbit in a homogeneous sample. This definition is supported by the representation of all the solar-mass binaries in any cluster of known age in a diagram orbital period vs age. However, the observational constraints on the tidal parameters, such as A and g of Eq.1 giving the circularization timescale, remain weak.

We have not examined in this paper the alternate tidal solution proposed by Tassoul (e.g. [Tas88]. During the writing of this paper, Dr. Tassoul kindly communicated us two important preprints in this context ([TaT92a], [TaT92b]) that we plan to take into account in a further paper.

Acknowledgements. We thank R.D. Mathieu for critical reading of the manuscript.

References

[AbL76] Abt H.A., Levy S.G. *Multiplicity among solar-type stars.* 1976, ApJS 30, 273-306

[AEE85] Abraham de Epstein A.E., Epstein I. *Photometric study of Zeta Sculptoris cluster.* 1985, AJ 90, 1211-1223

[AKa70] Artjukhina N.M., Kalinina E. *Proper motions of stars in the wide surrounding of the Pleiades.* 1970, Trudy Gos. Astron. Inst. Shternberga XXXIX, 111-190

[ArK69] Argue A.N., Kenworthy C.M. *Membership of the Coma star cluster.* 1969, MNRAS 146, 479-488

[Art66a] Artjukhina N.M. *New members in the corona of Praesepe.* 1966a, Trudy Gos. Astron. Inst. Shternberga XXXIV, 181-222

[Art66b] Artjukhina N.M. *The proper motions of stars in the wide surrounding of Praesepe cluster and dimension of the corona of the cluster.* 1966b, Trudy Gos. Astron. Inst. Shternberga XXXV, 111-157

[BMP79] Baranne A., Mayor M., Poncet J.-L. *CORAVEL - A new tool for radial velocity measurements.* 1979, Vistas Astron. 23, 279-316

[BoJ88] Boffin H., Jorissen A. *Can a barium star be produced by wind accretion in a detached binary?* 1988, AA 205, 155-163

[Bos88] Boss A.P. *Binary stars: formation by fragmentation.* 1988, Comments Astrophys. 12, 169-190

[Bra76] Branch D. *On the multiplicity of solar-type stars.* 1976, ApJ 210, 392-394

[CHN69] Constantine S.M., Harris B.J., Nikoloff I. *Proper motions in the field of the open cluster NGC 6475 (M7).* 1969, Proc. ASA 1, 207-208

[Cla92] Clarke, C.J. *1992, this volume*

[DuM91] Duquennoy A., Mayor M. *Multiplicity among solar-type stars in the solar neighbourhood. II. Distribution of the orbital elements in an unbiased sample.* 1991, AA 248, 485

[Egg70] Eggen O.J. *A very young cluster with a moderate metal deficiency.* 1970, ApJ 161, 159-162

[Egg86] Eggleton P.P. *From wide to close binaries?* 1986, The Evolution of Galactic X-Ray Binaries, eds J. Truempler et al., Reidel, pp 87-105

[Egg92] Eggleton P.P. *1992, this volume*

[Eps68] Epstein I. *Four-color photoelectric photometry of two high latitude clusters Blanco 1 and Melotte 227.* 1968, AJ 73, 556-565

[Fra89] Francic S.P. *Mass function for eight nearby galactic clusters.* 1989, AJ 98, 888-925

[Fre80] Fresneau A. *Membership of the α Per cluster as determined by proper motion.* 1980, AJ 85, 66-70

[GGZ85] Griffin, R.F., Gunn J.E., Zimmerman B.A., Griffin R.E.M *Spectroscopic orbits for 16 more binaries in the Hyades field.* AJ 90, 609-642

[Gli69] Gliese W. *Catalogue of nearby stars.* 1969, Veröff. Astron. Rechen Inst. Heidelberg, Vol.22

[GoM91] Goldman I., Mazeh T. *On the orbital circularization of close binaries.* 1991, ApJ 376, 260-265

[Gri67] Griffin, R.F. *A photoelectric radial-velocity spectrometer.* ApJ 148, 465-476

[GrG78] Griffin, R.F., Gunn J.E. *Spectroscopic orbits for the Hyades dwarfs van Bueren 62, 117, and 121.* 1978, AJ 83, 1114-1118

[GrG81] Griffin R.F., Gunn J.E. *Spectroscopic orbits for the Hyades dwarfs BD +23.635, vB 162, and vB 182, and the nonmember J318.* 1981, AJ 86, 588-595

[GMG82] Griffin R.F., Mayor M., Gunn J.E. *Spectroscopic orbits for three double-lined binaries in the Hyades field, 22.669, vA 771, and vB 166.* 1982, AA 106, 221-248

[Hal87] Halbwachs J.-L. *Distribution of mass ratios in spectroscopic binaries.* 1987, AA 183, 234-240

[Hei26] Heinemann K. *Photographische Photometrierung und Vermessung des Haufens NGC 752.* 1926, Astr. Nach. 227, 193-219

[Her47] Hertzsprung E. *Catalogue de 3259 étoiles dans les Pléiades.* 1947, Ann. Sterrewarte Leiden 19 No 1A

[HJI61] Hoag A.A., Johnson H.L., Iriarte B., Mitchell R.I., Hallam K.L., Sharpless S. *Photometry of stars in galactic cluster fields.* 1961, Publ. US. Nav. Obs. 17, 347-542

[Hof82] Hoffleit D., Jaschek C. *Catalogue of Bright Stars, Third Revised Version.* 1964, Yale University Observatory, New Haven.

[JaM88] Jasniewicz G., Mayor M. *Radial velocity measurements of a sample of northern metal-deficient stars.* 1988, AA 203, 329-340

[JoB92] Jorissen A., Boffin H. *Evidence for interaction among wide binary systems: to Ba or not to Ba? 1992, this volume*

[Joh53] Johnson H.L. *Magnitudes and colors in NGC 752. 1953, ApJ 117, 356-360*

[KW27] Klein-Wassink W.J. *The proper motion and the distance of the Praesepe cluster. 1927, Publ. Kapteyn Astr. Lab. 41*

[Koe59] Koelbloed D. *Three-colour photometry of the three southern open clusters NGC 3532, 6475 (M7) and 6124. 1959, Bull. Astr. Inst. Netherl. 14, 265-278*

[Kra65] Kraft R.P. *Studies of stellar rotation. I. Comparison of rotational velocities in the Hyades and Coma clusters. 1965, ApJ 142, 681-702*

[Lat85] Latham D.W. *Digital stellar speedometry. In "Stellar Radial Velocities", IAU Coll.88 (1985), eds A.G. Davis Phillip and D.W. Latham (Davis Press), p.21-34*

[Lav61] Lavdovski V.V. *A catalogue of proper motion of stars in 13 open clusters and in their vicinity. 1961, Trudy Glavn. Astr. Obs. 73, 1-131*

[LeH83] McLean B.J., Hilditch R.W. *Radial velocities for contact binaries - II. TZ Boo, XY Boo, TX Cnc, RZ Com, CC Com, and Y Sex. 1983, MNRAS 203, 1-8*

[LMC88] Latham D.W., Mazeh T., Carney B.W., McCrosky R.E, Stefanik R.P., Davis R.J. *A survey of proper motion stars, VI: Orbits for 40 spectroscopic binaries. 1988 AJ 96, 567-587*

[LMM92] Latham D.W., Milone, A.E., Mathieu, R.D., Davis, R.J. *Spectroscopic binaries in the open cluster M67. 1992, this volume*

[LTS92] Latham D.W., Torres, G., Stefanik R.P., Mazeh T., Carney B.W., Davis R.J. *Spectroscopic binaries in the halo. 1992, this volume*

[LuA92] Lubow S.H., Artymowicz P. *Eccentricity evolution of a binary embedded in a disc. 1992, this volume*

[LuW81] DeLuca E.E., Weis E.W. *A search for red-dwarf members of the Coma star cluster. 1981, PASP 93, 32-34*

[MaG92] Mazeh T., Goldberg D. *A new algorithm to derive the mass-ratio distribution of spectroscopic binaries. 1992, this volume*

[MaM81] Maeder A., Mermilliod J.-C. *The extent of mixing in stellar interiors: evolutionary models and tests based on the HR diagrams of 34 open clusters. 1981, AA 93, 136-149*

[MaM84] Mayor M., Mermilliod J.-C. *Orbit circularization time in binary stellar systems*. In "Observational tests of the stellar evolution theory", 1984, eds A. Maeder and A. Renzini, Reidel, Dordrecht, p.411-414

[MaM88] Mathieu R.D., Mazeh T. *The circularized binaries in open clusters: a new clock for age determination*. 1988, AJ 326, 256-264

[MaV91] Maceroni C., Van't Veer F. *The evolution and synchronization of angular-momentum-losing G-type main-sequence binaries*. 1991, AA 246, 91-98

[MaS79] Mazeh T., Shaham J. *The orbital evolution of close triple systems: the binary eccentricity*. 1979, AA 77, 145-151

[Mal27] Malmquist K.G. *Investigation of the stars in high galactic lattitude I. Lunds Medd Ser II no 37*

[Mal27] Malmquist K.G. *Investigation of the stars in high galactic lattitude II. Stockholms Obs. Ann. 12 no 7*

[Mat89] Mathieu R.D. *Spectroscopic binaries among low-mass pre-main sequence binaries*. 1989, Highlights in Astronomy, Vol.8, 111-115

[Mat92] Mathieu R.D. *The eccentricity distribution of pre-main sequence binaries*. 1992, this volume

[Maz90] Mazeh T. *Eccentric orbits in samples of circularized binary systems: the fingerprint of a third star*. 1990, AJ 99, 675-677

[MDL92] Mathieu R.D., Duquennoy A., Latham D.W., Mayor M., Mermilliod J.-C. *The distribution of cutoff periods with age: an observational constraint on the circularization theory*. 1992, this volume

[Mer88] Mermilliod J.-C. *Description of a database for stars in open clusters*. 1988, Inform. Bull. CDS 35, 77-91

[Mer92] Mermilliod J.-C. *The database for stars in open clusters. II. A progress report on the introduction of new data*. 1992, Inform. Bull. CDS 40, 115-120

[MeM85] Mermilliod J.-C., Mayor M. *Radial velocities of F and G dwarfs in the Orion cluster region*. 1985, Stellar Radial velocities, eds A.G.D. Philip and D.W.Latham (Davis Press), Schenectady, 367-370

[MeM92] Mermilliod J.-C., Mayor M. *Distribution of orbital element for red giants in open clusters*. 1992, this volume

[MGD92] Mazeh T., Goldberg D., Duquennoy A., Mayor M. *On the mass-ratio distribution of spectroscopic binaries with solar-type primaries*. 1992, ApJ (submitted)

[Mil91] Milone A.A.E. *Los blue stragglers de M67 y otros cumulos abiertos.* 1991, PhD Thesis, Córdoba

[Mit60] Mitchell R.I. *Photometry of the α Persei cluster.* 1960, ApJ 132, 68-75

[MLG90] Mathieu R.D., Latham D.W., Griffin R.F. *Orbits of 22 spectroscopic binaries in the open cluster M67.* 1990, AJ 100, 1859-1881

[MMM92] Meynet G., Mermilliod J.-C., Maeder A. 1992, AAS (submitted)

[MoG87] Morbey C.L., Griffin R.F. *On the reality of certain spectroscopic orbits.* 1987, AJ 317, 343-352

[MRD92] Mermilliod J.-C., Rosvick J., Duquennoy A., Mayor M. *Investigation of the Pleiades cluster. II. Binary stars in the F5-K0 spectral region.* 1992 AA (submitted)

[MWD90] Mermilliod J.-C., Weis E.W., Duquennoy A., Mayor M. *Investigation of the Praesepe cluster.I. Identification of halo members.* 1990, AA 235, 114-130

[NoD92] North P., Duquennoy A. *Are barium dwarfs progenitors of barium giants?* 1992, this volume

[Ols84] Olsen E.H. 1984, private comm.

[Pla91] Platais I. *The studies of proper motions in the regions of open clusters. II. NGC 752.* 1991, AAS 87, 69-87

[Pri89] Pringle, J.E. *On the formation of binary stars.* 1989, MNRAS 239, 361-370

[RoV61] Rohlfs K., Vanysek V. *Photometry of the galactic cluster NGC 752.* 1961, Astron. Abh. Hamb. Sternw. V no 11

[RMM92] Rosvick J.M., Mermilliod J.-C., Mayor M. *Investigation of the Pleiades cluster.I. Radial velocities of corona stars.* 1992, AA 255, 130-138

[SBB83] Smith M. A., Beckers J. M., Barden S. C. *Rotation among Orion Ic G stars: angular momentum loss considerations in pre-main-sequence stars.* 1983, ApJ 271, 237-254

[SHB85] Stauffer J.R., Hartmann L.W., Burnham J.N., Jones B.F. *Evolution of low-mass stars in the α Persei cluster.* 1985, ApJ 289, 247-261

[SHJ89] Stauffer J., Hartmann L.W., Jones B.F., McNamara B.R. *Pre-main-sequence stars in the young cluster IC 2391.* 1989, ApJ 342, 285-294

[SSM92] Schaller G., Schaerer D., Meynet G., Maeder A. *New grids of stellar models from 0.8 to 120 $M_\odot$ at Z=0.020 and Z=0.001.* 1992, AAS, in press.

[StL85] Stefanik R.P., Latham D.W. *The Hyades: Membership and convergent point from radial velocities*. in IAU Coll 88 Stellar Radial Velocities, Eds A.G.D. Philip and D.W. Latham, (Davis Press, Schenectady), p. 213-222

[Tas88] Tassoul, J.-L. *On orbital circularization in detached close binaries*. 1988, ApJ 324, L71-L73

[TaT92a] Tassoul, M., Tassoul, J.-L. *On the efficiency of Eckman pumping for synchronization in close binaries*. 1992, ApJ, in press

[TaT92b] Tassoul, J.-L., Tassoul, M. *A comparative study of synchronization and circularization in close binaries*. 1992, ApJ, in press

[TLM92] Torres G., Latham D.W., Mazeh T., Carney B.W., Stefanik R.P., Davis R.J., Laird J.B. *Tidal circularization among the close binaries in the halo*. 1992, to appear in the Proc. IAU Symposium No.151, Ed. R.S. Polidan, Dordrecht

[Tri90] Trimble, V. *The distributions of binary system mass ratios: a less biased sample*. 1990, MNRAS 242, 79-87

[Tru38] Trumpler R.I. *The star cluster in Coma Berenices*. 1938, Lick Obs. Bull. 18, 167-195

[TRJ89] Trullols E., Rossello G., Jordi C, Lahulla F. *uvbyβ photometry of 67 stars in the region of α Persei*. 1989, AAS 81, 47-50

[TuY87] Tutukov A.V., Yungelson L.R. *Merger of components in intermediate mass close binaries*. 1987, Comments Astrophys. 12, 51-65

[VeM88] Van't Veer F., Maceroni C. *The angular momentum loss of rapidly rotating late-type main-sequence binaries*. 1988, AA 199, 183-190

[VeM92] Van't Veer F., Maceroni C. *The dynamical evolution of G-type main-sequence binaries*. 1992, this volume

[vLA86] van Leeuwen F., Alphenaar P., Brand J. *A VBLUW photometric survey of the Pleiades cluster*. 1986, AAS 65, 309-347

[WGL88] Westerlund B.E., Garnier R., Lundgren K., Pettersson B., Breysacher J. *UBV and uvby-beta photometry of stars in the region of the Zeta Sculptoris cluster*. 1988 AAS 76, 101-120

[Wil76] Wilson R.E. *TX Cnc - Which component is hotter?* 1976, AA 48, 349-357

[ZaB89] Zahn J.-P., Bouchet L. *Tidal evolution of close binary stars, II: Orbital circularization of late-type binaries*. 1989, AA 223, 112-118

[Zah77] Zahn J.-P. *Tidal friction in close binary stars*. 1977 AA 57, 383-394.

[Zah89] Zahn J.-P. *Tidal evolution of close binary stars, I: Revisiting the theory of equilibrium tide. 1989, AA 220, 112-116*

Tidal Circularization of Short Period Binaries

Itzhak Goldman and Tsevi Mazeh *

Abstract

A modified approach to the tidal circularization of short-period binaries with convective envelopes is presented. It accounts for the reduction of the stellar viscosity expected when the tidal shear varies on a timescale which is comparable to or shorter than the typical convection turnover timescale. The above reduction depends on the value of the orbital period. As a result, in the present model, the circularization timescale for close binaries is proportional to the binary period to the power of 10/3. This exponent is smaller than in previously suggested theories. We briefly discuss the relevance of our model to the issue of main-sequence versus pre-main-sequence circularization.

1. Introduction

The availability of large samples of spectroscopic binaries has prompted recently a renewed interest in the orbital evolution of close binary systems. Specifically, the accumulating data indicate that short-period binaries are circular, while the longer-period binaries display a wide distribution of eccentricities. Consequently, the theory of tidal circularization was brought into focus. Different theoretical approaches have been suggested to estimate the circularization timescale and its dependence on the binary period.

It is clear that the circularization of a binary orbit is the result of the tidal interaction between the components of the system. The tidal deformation of the stars combined with stellar viscosity generate torques which tend to circularize the binary orbit. The first approximate calculations indicated that, for late-type stars with convective envelopes, the tidal interaction through the stellar main-sequence phase is strong enough to circularize the short-period binaries (Zahn [Zah66], [Zah77]; Alexander [Ale73]; Lecar, Wheeler, & McKee [LWK76]; [Hut81]). The circularization induced by the convective viscosity is characterized by a timescale which is proportional to the orbital period of the binary to the power 16/3 ([Zah77]; Mathieu & Mazeh [MMz88]).

*School of Physics and Astronomy, Sackler Faculty of Exact Sciences, Tel Aviv University, Tel Aviv 69978, Israel

A completely different mechanism for the circularization of short-period binaries has been suggested by Tassoul [Tas88]. It involves large-scale transient meridional currents induced by the tidal distortion of the stellar axial symmetry. Originally suggested to account for the synchronization or pseudo-synchronization observed in early-type stars with radiative envelopes [Tas87], the model was further applied to late-type binaries [Tas88]. The author claimed that this novel mechanism is more efficient than the one suggested by the previous works (see also Tassoul & Tassoul [TaT90]). Within this theory, the circularization timescale is proportional to the period of the binary system to the power 49/12.

Zahn [Zah89] pointed out that the variation of the tidal stellar deformation occurs in short-period binaries on a timescale comparable to or shorter than a typical stellar convective turnover timescale. This effect tends to make the convective viscosity less effective and makes the circularization timescale longer. Zahn & Bouchet [ZaB89] followed this argument and claimed that for binary systems with masses ranging from 0.5 to 1.25 $M_\odot$ the observed circular orbits could not have been circularized on the main-sequence. They argued therefore that the circular orbits are a result of circularization during the pre-main-sequence phase.

In the present paper we discuss a modified model for the circularization timescale, which takes into account the reduction of the effective viscosity in short-period binaries. In our model, the circularization timescale can be approximated as a new power law of the binary period. The new exponent, 10/3, is substantially smaller than that of the early theory of Zahn, and is even smaller than the exponent derived by Tassoul [Tas88].

In what follows we focus on the qualitative aspects of the model. For a more detailed and formal treatment see Goldman & Mazeh [GoM91], on which the present work is based.

2. The Modified Model for Tidal Circularization

The tidal interaction between two components of a stellar system induces a velocity gradient in the envelopes of the two stars. The viscosity in the envelopes causes a frictional force between adjacent fluid layers of different velocities. This in turn causes the induced tidal bulges to lag (or precede) the line connecting the centers of the two stars. The resulting torques lead to the synchronization of stellar rotation with the orbital period and to the circularization of the binary orbit. The timescales for these processes are *inversely proportional to the stellar viscosity*. Therefore, in any theory of circularization, the estimation of the viscosity is of prime importance. The basic source for viscosity— molecular viscosity, is due to transfer of momentum between fluid layers separated by a distance comparable to the molecular mean free path. The molecular kinematic viscosity is $\nu = \frac{1}{3}lv_{th}$, where l is the mean free path and v_{th} is the thermal velocity of the molecules (see, e.g., Reif [Rei65]). However, the molecular viscosity turns out to be too small to produce efficient circularization and synchronization. The effective viscosity can be largely enhanced if the stellar envelopes have a turbulent velocity field. This is indeed the case for stars with convective envelopes.

A fully developed turbulence (see, e.g., Batchelor [Bat73]; Hinze [Hin75]) can be viewed as a fluctuating velocity field composed of components with different spatial scales. Each such component can be (heuristically) visualized as a turbulent eddy which rolls over its size l with a velocity $v(l)$. The eddies interact among themselves, in a mode that can be described in terms of the breakup of an eddy into smaller eddies, resulting in a transfer of energy from the large scale eddies to the small scale ones. The timescale for this energy cascade is of the order of a roll-over time of an eddy $\tau(l) \sim l/v(l)$. It can be viewed as the lifetime of an eddy before it breaks up into smaller eddies. Typically, this timescale is longer for larger eddies. The energy cascade terminates at some small scale for which dissipation by molecular viscosity takes over. For the turbulence to be in a steady state, there must exist some stirring mechanism that continuously generates the large scale eddies and thus supplies the energy cascaded into the small scales.

Consider a nonturbulent velocity gradient imposed onto a turbulent medium, like the one induced by the tidal interaction in the convective stellar envelopes. An eddy can transfer momentum between fluid layers of different velocities (owing to the velocity gradient) separated by a distance comparable to its size. Thus, the turbulent velocity field which interacts with the velocity gradient produces a drag contributed by all the eddies. The contribution of the largest eddies is the dominant one, and therefore the effective turbulent viscosity is given by (see [GoM91])

$$\nu_t \sim v_0 l_0, \tag{1}$$

where l_0 is the size of the largest eddy and v_0 is its velocity.

The energy drained from the velocity gradient into the turbulence is transferred to smaller scales by the turbulence cascade and is ultimately dissipated at the smallest scales by the molecular viscosity. We have shown [GoM91] that for the case under consideration here, the rate of energy drained into the turbulent convection from the tidal velocity gradient is much smaller than the original rate of energy cascade; thus the turbulence is essentially unchanged by the interaction with the tidal velocity gradient.

Consider now a close binary with a small eccentricity in synchronous rotation. In this case the tidal velocity gradient varies around a mean value of zero on a timescale of the orbital period P. The above description of the interaction between the turbulence and the tidal velocity is valid when the timescale for variation of the tidal velocity gradient is much larger than τ_0, the timescale for the interaction of the largest eddy with the velocity gradient. However, for some binaries the orbital period P is comparable to or shorter than τ_0. Thus, over a timescale comparable to or longer than τ_0, the large eddies will effectively interact with a time *average* of the tidal velocity gradient, which equals zero. Thus, one expects that the effective turbulent viscosity exerted on the velocity gradient by the eddies will be significantly reduced. Zahn [Zah89] was the first to realize that such a reduction is relevant for the problem of circularization of close binaries, since the orbital periods are indeed comparable to the interaction timescale of the largest convective eddies in stellar envelopes.

Zahn [Zah89] proposed (following [Zah66]) that in binaries with period $P \leq \tau_0$, the largest eddies interact with the tidal velocity gradient on a timescale $P/2$ rather than τ_0. Thus, the effective scale over which the largest eddies can interact with the tidal velocity gradient is reduced from $v_0\tau_o$ to $v_0 P/2$. Consequently, Equation (1) implies that the reduction in the turbulent viscosity is $P/(2\tau_0)$. Goldreich & Keeley [GoK77] proposed a different approach in the context of the damping of solar pulsations by turbulent convection. They suggested that when the pulsation period P_p is smaller than τ_0, only those eddies with a lifetime smaller than P_p would contribute to the effective turbulent viscosity. The two approaches differ with regard to the largest eddies. While [Zah89] maintains that the largest eddies are still the dominant contributor to the interaction with the tidal velocity gradient, though less effectively, [GoK77] claim they do not contribute at all to the turbulent viscosity. For the particular case of turbulent convection with a Kolmogorov spectrum, the reduction obtained by [GoK77] is proportional to $(P_p/\tau_0)^2$.

We find the suggestion of Zahn ([Zah66], [Zah89]) unconvincing, since the largest eddy does not interact with the tidal velocity gradient before rolling over a distance comparable to its size. Therefore, we agree with Goldreich & Keeley [GoK77] that the largest eddies will not contribute to the turbulent viscosity. The presence of the tidal velocity gradient does not change this fact since, as noted above, the turbulence in the present case is effectively unchanged by the energy it drains from the tidal velocity gradient.

To further clarify this point we use an analogy with molecular viscosity. Consider a velocity gradient that changes on a timescale P shorter than or equal to the molecular *mean* free time between collisions τ. According to the approach of [Zah89] one should take $P/2$ instead of τ as the effective time for interaction with the velocity gradient. However, it is important to remember that most of the molecules *will not suffer a collision* after a time shorter than τ. Only a fraction

$$(1 - e^{-P/(2\tau)}) \sim P/(2\tau) \qquad (2)$$

of the molecules will do so. As a result, the molecular viscosity will be reduced by a factor of $(P/(2\tau))^2$. A detailed calculation that considers the contribution from molecules with all possible values of the time between collisions (taking into account the probability distribution of this free time) yields the same result. Thus, if one is willing to consider a breakup of the largest eddy on a timescale shorter than τ_0, the probability for this to happen is small and *is itself* $\propto P/\tau_0$, yielding a reduction of the turbulent viscosity of the largest eddies $\propto (P/\tau_0)^2$.

Returning to turbulent convection, we may conclude that the large-eddy part of the turbulence spectrum *will not* contribute to the turbulent viscosity. Only small enough eddies with lifetime shorter than λP ($\lambda < 1$) will be effective. The specific functional dependence of the reduced turbulent viscosity on P is determined by the form of the turbulence spectrum. We have shown that for turbulent convection in solar-type stars, the spectrum is of the Kolmogorov form for eddies with lifetimes shorter than $\sim 0.5\tau_0$ [GoM91]. This yields for short period-binaries a reduced turbulent viscosity $\nu_{t,short}$,

$$\nu_t = \nu_{t,short} = A\nu_{t,0}\left(\lambda P/\tau_0\right)^2 \quad ; \quad \lambda P < \tau_0 \ , \tag{3a}$$

while for long orbital periods

$$\nu_t = \nu_{t,0} \quad ; \quad P >> \tau_0 \ , \tag{3b}$$

where we use $\nu_{t,0}$ to denote the nonreduced turbulent viscosity of Equation (1) and A to denote a dimensionless constant. We have shown that for turbulent convection $A \sim 3$ [GoM91].

Turn now to the implication of the reduced viscosity on the circularization timescale T_{circ} for short-period binaries. Since

$$T_{circ} \propto \nu_t^{-1} P^{16/3} \tag{4}$$

we get

$$T_{circ} \propto P^{10/3} \quad ; \quad \lambda P < \tau_0 \tag{5a}$$

and

$$T_{circ} \propto P^{16/3} \quad ; \quad P >> \tau_0 \ . \tag{5b}$$

Equation (3a) was used to express the reduced $\nu_{t,short}$ in terms of the nonreduced $\nu_{t,0}$. Note that even with the reduced turbulent viscosity, the dependence of the circularization timescale on the binary period is that of a power law, albeit with a smaller exponent.

Equations (5a) and (5b) describe the asymptotic dependences of T_{circ} on the binary period P. The detailed change from one behaviour to the other depends on the specific form of the turbulent spectrum for eddies of sizes intermediate between those of the largest eddies and those of the small eddies that have the Kolmogorov spectrum. For simplicity we represent the change by using the approximate form

$$T_{circ} = T_0(P/P_0)^{10/3} \quad ; \quad P < P_0 \qquad (6a)$$

and

$$T_{circ} = T_0(P/P_0)^{16/3} \quad ; \quad P > P_0 \quad , \qquad (6b)$$

which assumes a sharp discontinuous change from one exponent to the other. In reality, the transition is a continuous one. Here P_0 is the value for which the exponent changes from 10/3 to 16/3. In this representation the circularization timescale of a binary with period P_0 is the same for the two different power laws and is denoted by T_0. Figure 1 illustrates the approximation.

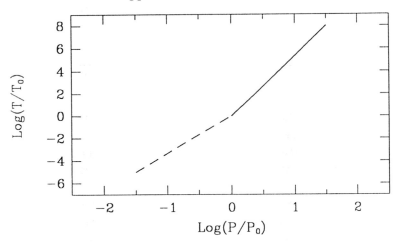

Figure 1. Schematic representation of the circularization timescale T_{circ} as function of the binary period P. *Solid line*: Equation (6b). *Dashed line*: Equation (6a).;

Up to this point we presented the theoretical basis of our model. The following sections examine the relevance of the theory to the observational data.

3.   Absolute Calibration of the Model

To calibrate the proposed model we present here an estimate of P_0 and T_0 for solar-type binaries. From Equations (3a) and (3b) we find

$$P_0 = \tau_0 \lambda^{-1} A^{-1/2}. \qquad (7)$$

The estimation of the convective timescale τ_0 is complicated by the fact that it is a function of depth in the convection layer. However, its weighted average will be close to the value at the lower part of the convection zone. This is so, since the tidal energy dissipated per unit volume is proportional to the gas density times the turbulent viscosity, both of which increase with depth. For solar-type stars the convective timescale at the lower part of the convective zone is estimated to be in the range 20–25 days in the model of Spruit [Spr74], while the corresponding value in the model of Goldreich & Keeley [GoK77] is $\simeq 12$ days. Thus, it seems plausible that τ_0 is in the range 10–25 days. The exact value of the parameter λ is not known but is likely to be in the range $1/2 - 1/(2\pi)$ where the limits correspond to the values proposed by [Zah89] and [GoK77], respectively. Application of the extreme values of the above parameters to Equation (7) yields values of P_0 in the range of 10–90 days. To be specific we arbitrary choose $\tau_0 = 17$ days and $\lambda=1/2$, so that Equation (7) yields

$$P_0 = 20 \text{ days.} \tag{8}$$

We note that from the convective model used by [Zah89] one can estimate the effective value of P_0 to be 18 days, in agreement with the above value.

The value of T_0 depends on the value of the nonreduced turbulent viscosity $\nu_{t,0}$; the estimation of the latter depends on the specific model used for the convection layers. In particular, the theoretical estimate depends on the ratio between the mixing length and the pressure scale height, and also on whether the model includes the possibility of overshooting of the convective eddies into the radiative stellar interior (VandenBerg & Poll [VaP89]). (Andersen, Nordström, & Clausen [ANC90] recently found the overshooting effect to be very important also in more massive stars with convective cores.) Different models yield different values for the depth of the convection zone and for the convective velocities, and consequently different values for the turbulent viscosity. In particular, incorporating overshooting into the model will increase the effective width of the convective zone and thus increase the turbulent viscosity and shorten the circularization timescale. Moreover, even a small amount of overshooting can be important, since the rate of energy dissipation per unit volume is proportional to the density, which is higher in the radiative interior. In view of these uncertainties we can only agree with the comment of Lecar *et al.* [LWK76] that the theoretical absolute calibration of the circularization timescale is uncertain by, at least, a factor of ten. Being aware of this situation we nevertheless wish to obtain an estimate for T_0.

We consider a binary consisting of two 1 $M_\odot$ stars and apply the the nonreduced turbulent viscosity from the convection model used by Zahn ([Zah89], see his Eq. [15]). For $P_0 = 20$ days and for the above nonreduced viscosity we find that

$$T_0 = 1.2 \times 10^{12} \text{ yr.} \tag{9}$$

We obtained a very similar value for T_0 when using the estimates of Mathieu & Mazeh ([MMz88], see their Table 1). The values of P_0 and T_0 should be used in Equations (6a) and (6b). Obviously, the theoretical values must be confronted with the observational data.

4.   Discussion

One way to use the observational data to constrain the different circularization theories is to consider coeval samples of binaries of known ages. One expects that in a given sample all the short-period binaries up to some transition period would be circularized. Of course, even in a strictly coeval sample of binaries, the transition might not be sharp, due to the fact that the stars can be different for different binaries. Nevertheless, transition periods have been found for five coeval samples with different ages—Pleiades $(0.1 \times 10^9$ yr), Hyades $(0.8 \times 10^9$ yr), the solar-type field stars $(4.5 \times 10^9$ yr), M67 $(5 \times 10^9$ yr), and the Galactic halo stars $(15 \times 10^9$ yr); see Mathieu et $al.$ ([MDL92], this volume). The corresponding transition periods are distributed between 5 and 20 days. Indeed, the detected transition periods are not sharp and many more binaries are needed to establish the exact value for each sample. In fact, even the definition of the transition period is not agreed upon. While Duquennoy, Mayor, & Mermilliod ([DMM92], this volume) suggested to identify it as the longest circularized period, Mazeh et $al.$ [MLM90] used the shortest eccentric binary period (for a discussion see [MDL92]). Nevertheless, regardless of the uncertainties in the transition periods and the ambiguity of their definition, the data seem to indicate an age dependence of the value of the transition period [MDL92]. The obvious interpretation is that circularization takes place during the main-sequence phase.

One of the transition periods determined recently is that of the Galactic halo stars (Latham et $al.$ [LMS92]). The quite long transition period—18.7 days, is of particular interest. For the values of P_0 and T_0 given by Equations (8) and (9), the circularization timescale corresponding to a period of 18.7 days is 9.6×10^{11} yr. This is 65 times larger than the estimated age of the halo! In order that the theoretical circularization timescale coincides with the observed age, our previous estimate of the nonreduced turbulent viscosity must be enhanced by the same factor. Even if one wished to use the power law of Equation (6b), the circularization timescale would be essentially the same, and would require a similar enhancement. We wish to stress that the need for an apparently large enhancement of the nonreduced turbulent viscosity is therefore a feature of any model of tidal circularization relying on convective turbulent viscosity. If indeed further observations will support the [LMS92] finding about the long transition period in the Galactic halo sample, then this exercise illustrates how the observational data can be used to calibrate the theoretical models.

Zahn & Bouchet [ZaB89] also noted that the theoretical estimate of the absolute calibration yields circularization timescales that are too long to allow for circularization on the main-sequence. They chose not to change the theoretical calibration but instead proposed that the circularization occurred during the pre-main-sequence (PMS) phase. However, such an assumption yields (in their model) transition periods of $\sim$ 5–8 days and could not explain the transition periods of M67, the solar-type field stars, and in particular the 18.7 days transition period of the Galactic halo sample.

Even though the data do not support their assumption that *all* circular orbits originated in the PMS phase, we regard as plausible the possibility that the transition periods in the range 5–8 days observed in young samples are indeed a result of PMS circularization. Thus, the transition period of the Hyades sample might indeed be the outcome of PMS circularization. Therefore, we suggest that only old enough, main-sequence samples are relevant for the testing of the different circularization models on the main-sequence. One should however bear in mind that it is not clear whether the simple tidal circularization approach applies for the PMS binaries since other effects (mass loss, mass transfer, magnetic torques) are expected to be important.

From the above discussion follows that the absolute calibration of the circularization timescale is of little use for the confrontation of different theoretical models with the observations. Instead, one can compare the observed transition periods with the different theoretical predictions regarding the functional dependence of the circularization timescale on the binary period. A reduction of the turbulent viscosity according to the approach of [Zah89] yields a power-law dependence with an exponent of 13/3, while our approach predicts an exponent of 10/3, and the nonreduced viscosity model has an exponent of 16/3. Therefore, the observational data can in principle determine which of the different circularization models is favored (Jasniewicz & Mayor [JaM88]; [MDL92]; [LMS92]).

The different models for circularization should be compared in the future with more extensive data. More main-sequence binaries are one source of observational importance. Another source is the chromospheric active binaries with short periods. As Hall & Henry [HaH90] have convincingly shown, these systems can also test the circularization and synchronization theories, despite the fact that the different stars in the sample have different ages. After finding the correct theory of circularization, we can turn the reasoning backward and use the observed transition period as an independent 'clock' to estimate the sample age, as suggested by Mathieu & Mazeh [MMz88]. This method can, in principle, be used to check stellar evolution models.

Acknowledgements.

We thank A. Duquennoy, D. Latham, R. Mathieu, M. Mayor, and J.-P. Zahn for very fruitful discussions. This work was supported by the U.S.-Israel Binational Science Foundation grant 90-00357, and the Fund for Basic Research at Tel Aviv University.

References

[Ale73] Alexander, M.E. *1973, Ap. Space Sci., 23, 459*

[ANC90] Andersen, J., Nordström, B., & Clausen, J.V. *1990, Ap. J. Lett., 363, L33*

[Bat73] Batchelor, G.K. *1973, The Theory of Homogeneous Turbulence (Cambridge: Cambridge University Press)*

[DMM92] Duquennoy, A., Mayor, M., & Mermilliod, J.-C. *1992, in Binaries as Tracers of Stellar Formation, eds. A. Duquennoy & M. Mayor (Cambridge: Cambridge University Press)*

[GoM91] Goldman, I., & Mazeh, T. *1991, Ap. J., 376, 260*

[GoK77] Goldreich, P., & Keeley, D.M. *1977, Ap. J., 211, 934*

[HaH90] Hall, D.S., & Henry G.W. *1990, in NATO Advanced Study Institute, Active Close Binaries, ed. C. Ibanoglu (Dordrecht: Kluwer) p. 287*

[Hin75] Hinze, J.O. *1975, Turbulence (New York: McGraw-Hill)*

[Hut81] Hut, P. *1981, Astr. Ap., 99, 126*

[JaM88] Jasniewicz, G., & Mayor, M. *1988, Astr. Ap., 203, 329*

[LMS92] Latham, D.W., Mazeh, T., Stefanik, R.P., Davis, R.J., Carney, B.W., Krymolowski, Y., Laird, J.B., & Morse, J.A. *1992, to be submitted to A. J.*

[LWK76] Lecar, M., Wheeler, J.C., & McKee, C.F. *1976, Ap. J., 205, 556*

[MMz88] Mathieu, R.D., & Mazeh, T. *1988, Ap. J., 326, 256*

[MDL92] Mathieu, R.D., Duquennoy, A., Latham, D.W., Mayor, M., Mazeh, T., & Mermilliod, J.-C. *1992, in Binaries as Tracers of Stellar Formation, eds. A. Duquennoy & M. Mayor (Cambridge: Cambridge University Press)*

[MLM90] Mazeh, T., Latham, D.W., Mathieu, R.D., & Carney, B.W. *1990, in NATO Advanced Study Institute, Active Close Binaries, ed. C. Ibanoglu (Dordrecht: Kluwer) p. 145*

[Rei65] Reif, F. *1965, Fundamentals of Statistical and Thermal Physics (New York: McGraw-Hill)*

[Spr74] Spruit, H.C. *1974, Solar Phys., 34, 277*

[Tas87] Tassoul, J.-L. *1987, Ap. J., 322, 856*

[Tas88] Tassoul, J.-L. *1988, Ap. J. Letters, 324, L71*

[TaT90] Tassoul, J.-L., & Tassoul, M. *1990, Ap. J., 359, 155*

[VaP89] VandenBerg, D.A., & Poll, H.E. *1989, A. J., 98, 1451*

[Zah66] Zahn, J.-P. *1966, Ann. Ap, 29, 489*

[Zah77] Zahn, J.-P. *1977, Astr. Ap., 57, 383*

[Zah89] Zahn, J.-P. *1989, Astr. Ap., 220, 112*

[ZaB89] Zahn, J.-P., & Bouchet, L. *1989, Astr. Ap., 223, 112*

Composite-spectrum Binaries

R. Elizabeth M. Griffin *

A composite–spectrum binary consists of a cool (G, K or M) giant or supergiant primary and an A or B dwarf companion; the combination of spectral energies is such that both spectra are visible, superimposed and confused, in the photographic UV (in the general vicinity of the Ca II H & K lines). This composite spectrum can yield the mass ratio q of the component stars directly, as the inverse of the ratio of the radial–velocity separations of their spectra. However, although the radial velocity of the primary can be measured accurately with a Coravel–type instrument, that of the dwarf can only rarely be measured too by that method; its (relatively few) spectral lines are usually badly blended with those of the primary, and it is not possible in practice either to measure or to classify its spectrum in the presence of that of the primary. It is therefore essential to separate the two spectra, and that can now be achieved by digital subtraction.

The principle of the subtraction technique is to select from a library of standards a close match to the spectrum of the primary, and to subtract just the correct fraction of that surrogate from the composite spectrum such that all primary–spectrum features disappear simultaneously. The residual spectrum, that of the secondary star, may then be classified quite accurately and measured for radial velocity. Fig. 1 shows the results of applying the technique to the composite system HR 6902. The details and difficulties of the method are discussed by Griffin [Gri86].

In favourable cases, where the radial–velocity amplitude is high and the secondary lines are sharp, the mass ratio may be determined as precisely as ±0.02 (e.g. HR 6902 [Gri86], but if – as is true of many B and A–type dwarfs – the secondary is rotating rapidly enough to wash out the metallic lines and the velocity has to be determined from the Balmer lines, themselves already very wide even in the spectra of dwarfs with very low rotational velocities, the precision with which q can be measured is reduced. Of course, such difficulties imposed by Nature adversely affect efforts by whatever method to determine the mass ratio of the components in a composite–spectrum binary.

In special cases where an atmospheric eclipse is in progress – as happens periodically in a ζ Aur system – a third spectrum, that of the giant's chromosphere (seen in absorption along the line of sight to the almost–eclipsed dwarf), is superimposed

*Cambridge Observatories, U.K.

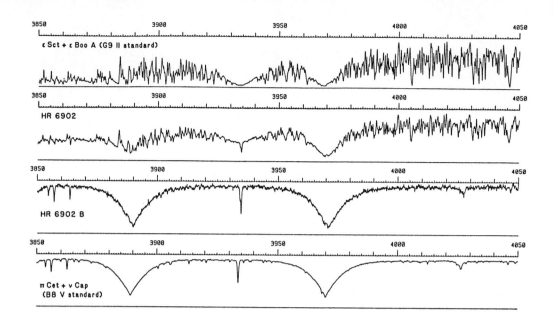

Figure 1. Subtraction of spectra. The top spectrum is adopted as the analogue of the late-type component of HR 6902. Subtraction of an optimal proportion of it from the composite spectrum (second tracing down) isolates the early-type component (third tracing), which may then be compared with the standard spectrum at the bottom. The horizontal line below each tracing is the zero-intensity line.

on the composite spectrum. Since many of the chromospheric lines are common to both the photospheric and the chromospheric spectra of the giant, the fidelity of the subtraction is of paramount importance. Nature may fortunately provide an opportunity to obtain the ideal spectrum for subtraction: that of the giant alone, if the eclipse is total. Uncovery of both spectra in the prototype system ζ Aur is demonstrated in Fig. 2; a full account is given by Griffin et al. [GGSR90].

The values of q determined for a number of composite binaries show considerable uniformity. Since the systems are sufficiently wide to be not interacting, the picture which is emerging represents a cross–section of stellar evolution: the two components are presumably coeval but one has evolved away from the main–sequence and the other has not. Most of the mass–ratios are close to 1.2 or 1.3; the tightness may be reflecting the fact that stars do conform fairly rigorously to a mass–luminosity relationship, while the specific value merely indicates those masses whose corresponding luminosities are comparable at λ 4000 Å. A few systems appear to have mass ratios that are definitely smaller, and may be cases in which mass has been transferred from the initially more massive star onto its companion.

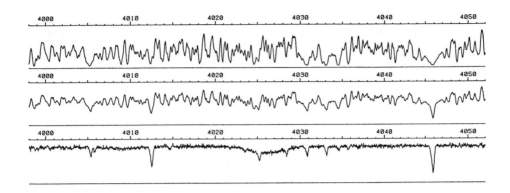

Figure 2. The special case of an atmospheric eclipse. The top panel shows the spectrum of the giant component of ζ Aur, as observed when it was totally eclipsing its B-type companion during November 1987. The second panel shows the composite spectrum with superimposed chromospheric absorption features, as seen about 0.75 day before first contact. Subtraction of a suitable proportion of the total-eclipse spectrum uncovers not only the broad lines of the dwarf (bottom panel; He I λ 4009 and 4026 Å can both be seen here) but also a forest of narrow features that are caused by absorption of B-star radiation within the intervening chromosphere of the giant. Since the chromosphere is not homogeneous, the contribution of chromospheric features to the uncovered B-star spectrum will depend upon the relative configuration of the two stars at the time of the observation.

References

[Gri86]   Griffin, R. & R., 1986, *JA&A* **7**, *195*

[GGSR90] Griffin, R.E.M., Griffin, R.F., Schröder, K.-P., Reimers, D., 1990,   *A&A* **234**, *284*

Orbital Elements for Field Late-Type Binaries

Roger F. Griffin *

My talk this afternoon will be rather clearly divided (like Gaul) into three parts, and instead of apologizing for giving such a scrappy presentation I am making the excuse that some slight connection might be traceable between the sections! First, in view of the ostensible reason for holding this Workshop and the fact that Paper 100 is not actually in print yet – though I assure you that it is duly in press in the October issue of *The Observatory* – I feel that I should describe that paper to you. Secondly, leading on from there, it seems useful to review the characteristics of the whole ensemble of the hundred orbits. And thirdly, I shall present a sort of obituary of the original radial-velocity spectrometer which has provided most of the data for the orbits.

The most recently published of the orbit papers in *Observatory Magazine*, Paper 99, concerns the fourth-magnitude K-giant star ϕ Piscium. The paper does mention that that star possesses a tenth-magnitude visual companion about eight seconds of arc away, but it is uncharacteristically reticent about it. That is because Paper 100 is devoted to that companion, which itself has turned out to be a single-lined spectroscopic binary whose nature is as amusing when understood as it was perplexing previously.

Of course it does create a good deal of observational difficulty when the object of interest is accompanied at such close range in the sky by one that is a hundred times brighter. For that reason the faint star, ϕ Piscium B, was initially observed only at Palomar – that was in the days, now unfortunately past, when Jim Gunn and I used to enjoy regular observing runs with the radial-velocity spectrometer that we built in 1971 for the coudé focus of the 200-inch reflector. With such a vast telescope and often with good or goodish seeing, even the companion star was seen as quite a bright object in its own right, and there was usually little difficulty in observing it despite the proximity of the almost dazzling primary. It was evidently a tolerably late-type star, because it did give a 'dip' on the radial-velocity traces. The dip was wide, showing the star to be in rapid rotation which suggested membership in a short-period binary system. As the results accumulated, the velocity was found at different times to be far above, or far below, the gamma-velocity of the primary star, a gamma that the secondary could be expected more or less to share. The desire to

---

*The Observatories, Madingley Road, Cambridge CB3 0HA, England

elucidate the orbit prompted extreme efforts at Cambridge to observe the faint star, which was in truth almost beyond the reach of the old spectrometer there. Even after prolonged scanning, I was never entirely happy that I had measured the shallow and diffuse dip that the star was known to give, rather than a slight dip that arose from scattered light from the all-too-adjacent primary. My misgivings were increased when I found that Cambridge observations always gave velocities relatively close to the velocity of the primary star – never far removed from it like the unassailable Palomar measurements.

Finally there came a time when curiosity over this seeming dichotomy of results got the better of me: ϕ Piscium B was treated as an object of the utmost priority at Palomar and measured repeatedly on the four nights of the annual observing run that was nominally devoted to the Hyades. On alternate nights the velocity proved to be either high and rising or low and falling, so the period was evidently close to two days. More exact calculations then readily disclosed the phasing of all the observations that had been made in previous seasons, and brought to light the amazingly close coincidence – to within about five seconds – between the period and the exact value of two sidereal days. The apparent disagreement between Palomar and Cambridge was immediately resolved: from any one place one could only observe the system in the vicinities of one or other of two particular phases, and whereas at Palomar those phases were approaching the nodes of the orbit, at Cambridge they were just at the conjunctions! – the Cambridge velocities were mostly quite reasonable after all!

Turning now to the second item of this talk, I should like to recall to you that the orbit series in *Observatory* included, after Paper 50, a synopsis of the first fifty orbits. Comparisons were made between the period and amplitude distributions of those fifty and of the orbits of the late-type binaries in Batten et al.'s *Seventh Catalogue*, which fairly represented the state of play at about the time that the *Observatory* series of orbits began. The comparisons were portrayed as histograms with 'bins' half a logarithmic unit wide, and they showed really dramatic differences in character between my orbits and the *Catalogue* ones.

The maximum frequency of periods in the *Catalogue* orbits came in the bin covering the range from three-and-a-bit to ten days, whereas the maximum for the *Observatory* series was in the bin from 300-odd to a thousand days. An analogous disparity was apparent in the amplitude histograms, but it was quantitatively smaller and of course of opposite sign – not surprising when one remembers that Kepler's Laws show that the amplitude goes as the inverse cube root of the period, other things being equal. The origin of the gross disparities in the characteristics of the two sets of orbits was ascribed principally to observational selection: few previous observers had been able and willing to follow small-amplitude binaries around lengthy orbital cycles. Indeed, it was pointed out that, even within the *Observatory* series, the effects of the steadily increasing time base were clearly apparent as a secular trend towards longer periods.

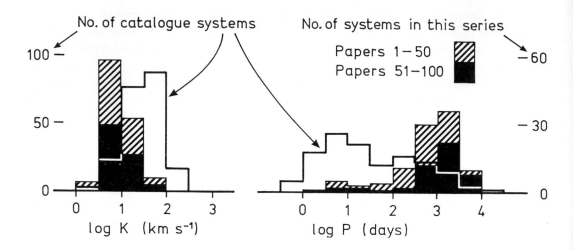

Figure 1. The distributions of the amplitudes (left) and the periods (right) of the orbits published in the *Observatory* series of papers, in comparison with those listed in Batten et al.'s *Seventh Catalogue*. The *Catalogue* sample has been restricted to systems with orbits of quality *c* or better and with at least one component later than type F5.

In an updated synopsis, now embracing the complete hundred orbits, the same trend is seen to continue (Fig. 1). The maximum frequency in the period histogram has moved up into the 1000-to-3000-day bin, and there is now a substantial population in the succeeding bin, representing orbits (such as that of ϕ Piscium A) with periods between 3000-odd and 10000 days. Although it seems obvious that selection effects are less extreme in the *Observatory* orbits than in the *Catalogue* ones, they certainly remain considerable, and will inevitably continue to do so as long as fresh objects are being added to the observing programme: those that turn out to have short periods will be documented before the long-period ones have had time to complete a single cycle. It is possible to remedy that particular statistical shortcoming to some extent, albeit at the cost of greatly reducing the size of the data pool, by considering only those binaries that were on the observing programme at some time so long ago that the periods of all of them are now known. But then it could be argued that short-period stars would be under-represented, because they had quickly been fully observed and written up and so had disappeared from the programme! The only reasonably safe time to take as a starting point in the present context seems to be the time just before the series of papers started, before any orbits had been written up and the corresponding stars dropped from the programme. Stars that were already under observation at that time do indeed show a still heavier weighting towards long periods than is seen in Fig. 1 – some of them are in fact still under observation seventeen years later.

Of course, none of this messing about with the statistics goes anywhere towards addressing the still more fundamental question concerning how the stars got onto the observing programme in the first place and the dreadful selection effects that must have operated there! The only thing I can usefully say about that is that the majority of the stars observed are ones that I myself have discovered to be binaries, and once a binary is discovered it is never dropped from the programme until its orbit is determined. Since the discovery procedure relies on seeing a discrepancy between the first few (usually only two or three) measurements, it clearly discriminates against small amplitudes. It can scarcely be doubted that a picture analogous to Fig. 1 but entirely freed from selection effects would show a still heavier emphasis on long periods and small amplitudes, with the trends extending well beyond the limits of the bins populated in the present histograms.

The new synopsis – it was submitted for publication for the same issue of *Observatory* as Paper 100 but has been held up by the referees – goes on to investigate two distributions that were neglected in its predecessor. One is the distribution of the longitudes of periastron. Many samples of binary orbits, including the *Catalogue* one, show a preponderance in the first quadrant; but mine do not, which is something of a relief, because the effect has been ascribed to spectroscopic anomalies related to gas streams which ought not to be present in the relatively wide binaries that constitute the great majority of my sample.

The more significant of the two newly plotted distributions is one relating eccentricity to orbital period – actually to the logarithm of the period, since otherwise most of the points are crowded into a ridiculously small part of the diagram. It is shown in Fig. 2. A great deal of interest has arisen in recent years in the matter of tidal circularization of orbits, and the Figure shows that most – but significantly enough not all – of the orbits with periods less than about ten days are circular, as they 'ought' to be. The diagram does more than just show signs of there being a critical period below which most orbits are circular: it shows a very obvious positive correlation, extending all the way to the longest periods, between eccentricity and log P. Since tidal effects are supposed to vary as a high power of the orbital separation, they would seem at first sight not to be able to play any part in the trend that one can see even at the right-hand side of the Figure. I would not want to be too dogmatic about that, though, because for tide-raising purposes the relevant quantity is not really the period but the separation of the binary components at periastron. If we used periastron separation instead of period as the abscissa of Fig. 2, the highest points would in general be the very ones that would be moved furthest to the left, so the trend would certainly be reduced and might even turn out to be removed. Periastron distances are proportional to $1/e$, so the most eccentric system, with e about 0.9, has a periastron radius reduced by a factor 10 compared with a circular orbit of the same period. Because of the three-halves power law between orbital radius and period, that high point in the Figure would in effect be moved $1\frac{1}{2}$ log units to the left. I think

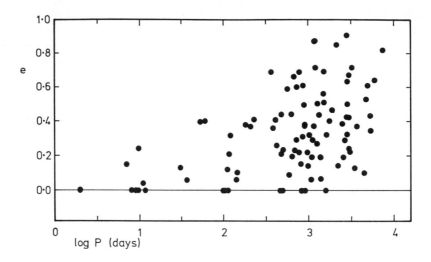

Figure 2. Distribution of the eccentricities of the 100 orbits as a function of log P.

it is easy to see qualitatively how a good sprinkling of the high eccentricities could be pushed well over towards the left of such a diagram. Unfortunately I cannot plot such a diagram for you, because the determination of periastron distances involves knowledge of both orbital inclinations and mass ratios, quantities that are usually unknown.

One thing that I *can* do that may be quite interesting in the context of the high incidence of binaries having long periods is to give you a preview of the radial-velocity curves of some binary systems with still longer periods than most of those so far published. Here I have viewgraphs showing preliminary orbits for such stars, relatively bright but largely neglected up till now by radial-velocity observers, as 14 Trianguli (period about 6200 days), ζ Cygni (6500 days), χ Andromedae (7600 days), κ Persei (10,400 days) and HR 4593 (probably about 12,000 days) (Fig. 3a).

Although the periods of those stars are long, they are known, at least approximately. There are quite a few other stars that have been under observation for comparable lengths of time, of the order of twenty years, and have shown definite changes of velocity but are nowhere near to completing an orbital cycle; in fact some of them look as if twenty years may be quite a small fraction of the period. I am illustrating the velocity changes of some of them in Fig. 4. One last object to which I will draw attention is HD 176695, a star that has shown a gradual rise in velocity ever since I started to watch it eighteen years ago. On the rather tenuous evidence of some old David Dunlap observations I think that its period will turn out to be about 7600 days; you can see from the diagram (Fig. 3b) why the immediate future promises

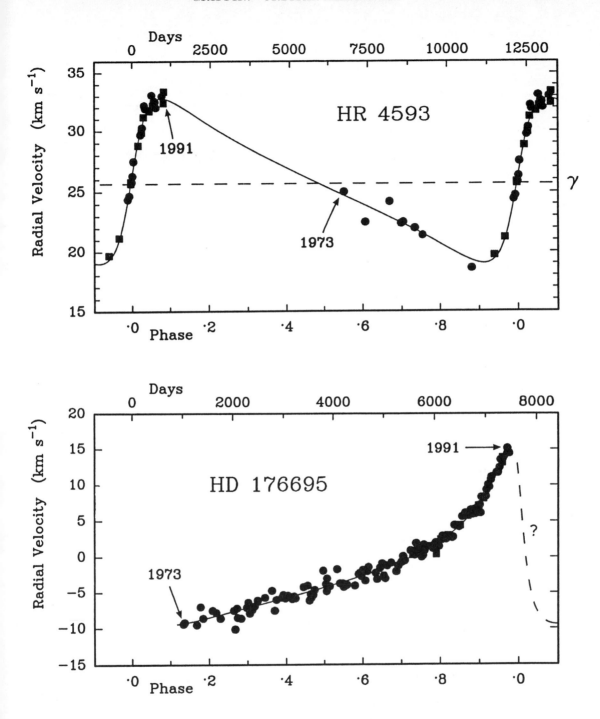

Figure 3. Preliminary radial-velocity curves for HR 4593 and HD 176695.

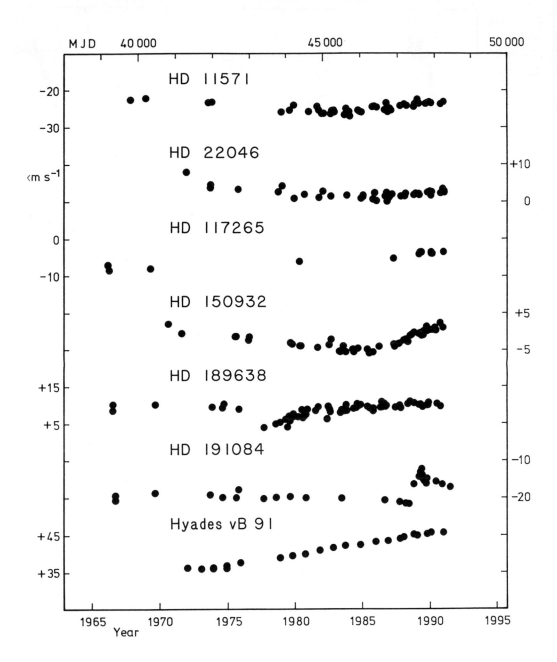

Figure 4. Radial-velocity observations of a number of slow spectroscopic binaries.

to be a significant time. Since at present I have no equipment at Cambridge I have asked the Coravel observers to watch it for me – you can imagine why I do not want to have to wait another cycle to cover this part of the orbit!

That brings me to the final part of my talk, which concerns the original radial-velocity spectrometer. The instrument has recently been dismantled; the business end of it is to go to the Science Museum in London, and the rest has been thrown in a skip! At the time of its development it seemed a very powerful instrument: in the early years it certainly measured radial velocities of a high standard far more quickly and easily than had been possible before, but it has long ago become hopelessly out of date with its chart-recorder output and the need to reduce its traces by bisecting the 'dips' by eye. A particular drawback in comparison with newer spectrometers was its inability to integrate the trace: one simply scanned backwards and forwards, quite slowly, and every time you passed through the dip the machine drew it for you in 'real time' – about thirty seconds. If you couldn't measure the velocity from thirty seconds' worth of trace you couldn't measure it at all. There was nothing to stop you going on scanning back and forth as long as you liked, but all you got were more traces of the same character, so there was no improvement whatever in the signal-to-noise ratio. In fact, the harder you tried to measure a star that gave an unreadable trace the more bother you gave yourself when you came to try to reduce the observation! So I have not been too sorry to see the last of the old instrument, especially as it is hoped soon to instal in its place a new one which is in essence a Coravel, for which the engineering drawings and many other data have most generously been made available by Michel Mayor and his collaborators.

It may not be out of place to mention here that difficulties of an instrumental nature were not the only ones that had to be overcome before the cross-correlation technique could begin to make significant contributions to the radial-velocity literature. In these days when the vast majority of velocities are measured by cross-correlation it may appear unbelievable that there could have been such seemingly organized resistance to it as I certainly encountered in my efforts to get the early papers published. The fact is that up until then radial velocities had been measured 'properly' on photographic spectra on a line-by-line basis, and experts in the field had amassed a great deal of empirical experience of how, at any given spectral type, certain lines should be scrupulously avoided because they were "bad for radial-velocity purposes". In plain language, that simply meant that they were blends, the relative strengths of whose contributors were changing rapidly in the vicinity of the relevant spectral type, so the wavelength of the blend was prone to vary from star to star. In a wholesale cross-correlation of the spectrum one did not bother with such niceties, nor was there any reason to do so: by integrating the result from so many lines, shifts arising from individual blends were averaged out to an extent that was statistically altogether acceptable. Of course I knew that, and tried to protect myself at the outset by rehearsing the argument quantitatively in the initial paper on the method, but it took a long time to persuade the Establishment of its validity. They seemed to find it particularly vexing that I was not even concerned about what spectral type I was measuring, so long as it gave a dip! Meanwhile, paper after paper was stalled for unconscionable times by anonymous referees – I had better not say too much in case they are here in person, but what I had thought of saying this afternoon was that what seemed to me to be the real trouble was that they couldn't bear to hear how easy it was to measure good velocities by the new technique!

Years passed, and although some interest was expressed in certain quarters concerning the method, nobody but my collaborators and I seemed to be using it; so in the end I decided that an advertising campaign was called for, a campaign to brandish the virtues of the method in front of the astronomical public at short and regular intervals. I had for some considerable time been taking an interest in spectroscopic binaries, and changes in emphasis at Cambridge following the retirement of Professor Redman gave me virtually unfettered use of the 36-inch telescope and enabled binaries to be watched on a beneficially regular basis. I accordingly decided to institute in *Observatory Magazine* (where I had the unfair advantage of being an Editor) the series of papers that we celebrate now, with the words "PHOTOELECTRIC RADIAL VELOCITIES" prominently displayed in the title. I accumulated enough data beforehand that I could feel reasonably confident that the series, once started, could be kept going indefinitely in every issue of the *Magazine* until kingdom come! Little did I realize then that, only two or three years later, Coravel would burst upon the scene and render any further advertisement unnecessary; but by the time that that had happened, the series was fairly under way and no useful purpose would be served by altering its title or character at that stage. From time to time, both before and after I retired from the Editorial Board of *The Observatory*, the Editors have had misgivings about the propriety of publishing all these papers; but the longer the series continues the more difficult it is to be seen to decide that it has all been a ghastly mistake, so the result of reconsideration has always been – touch wood! – that things may take their course for the time being, although the length of Paper 100 was almost the last straw! Anyway, that is how the series came to be started, and I hope that the Editors will see the attendance at this conference as constituting some measure of approval of their public spirit in publishing the papers!

Figure A. The author at the observer's desk.

The 36-inch telescope is seen in the background; the casing of the collimator arm of the spectrometer is at the left. When the observer is seated at the desk he has the telescope controls to his left, the eyepiece (to which the coudé field is re-imaged) conveniently in front of him, and the spectrometer controls and chart output to his right. The structure in the lower right corner of the picture, and the lead weight dangling ominously just above the author's head, are associated with the dome.

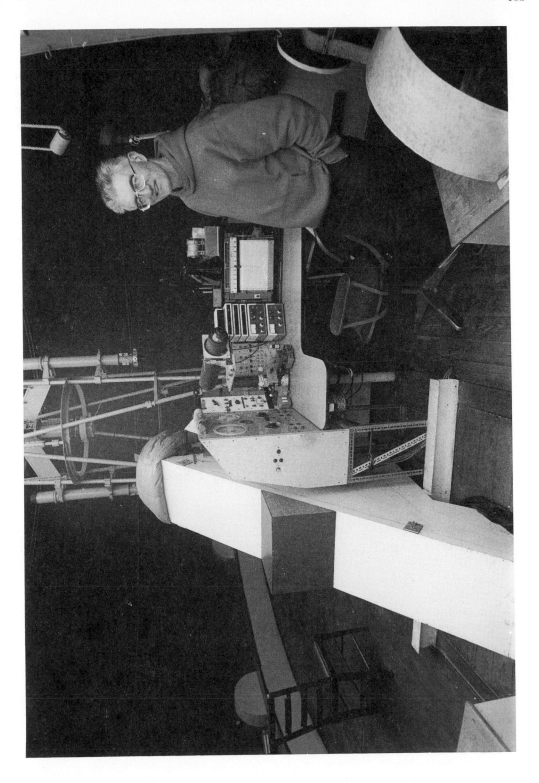

Figure B. The Cambridge radial-velocity spectrometer, seen from the west.

Light from the telescope (out of the picture to the right) travels northwards to the coudé focus at the entrance slit, which is within the rectangular hood at the top right of the instrument. Light that passes through the slit is intercepted by a small reflecting prism and sent towards the lower left; it is collimated and returns to a diffraction grating whose mounting is insulated from abrupt temperature changes by the conspicuous sleeping bag draped round it.

The diffracted beam goes vertically downwards through a camera lens to a focal-plane assembly below the floor level. The coudé field is re-imaged by the small lens seen below the slit hood, after reflection by the mirror immediately to its right, at the eyepiece through which the author is looking.

In the foreground are the electronics of the spectrometer, which employs three separate photomultipliers with independent power supplies, and the strip-chart ('Brown') recorder which draws the radial-velocity trace. On top of the Brown Recorder is a gearbox providing a choice of scanning speeds.

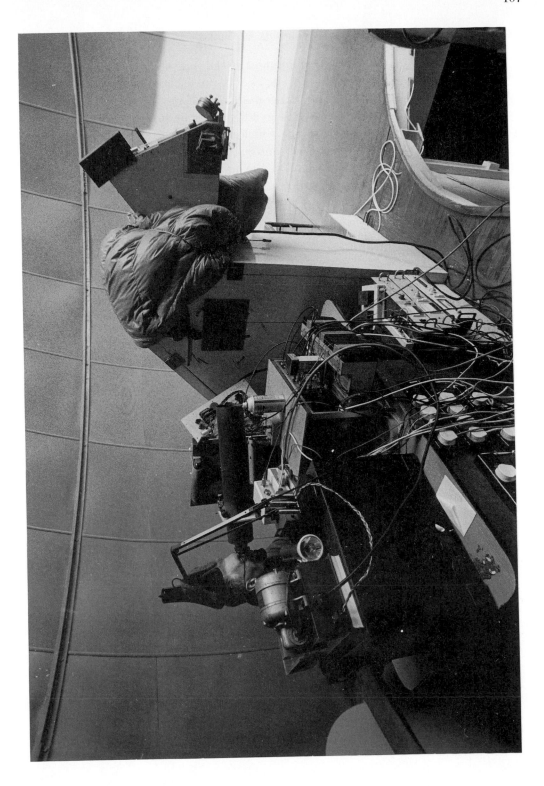

Figure C. The focal-surface assembly of the Cambridge spectrometer.

This was the novel part of the equipment at the time of its inception, and has now gone to the Science Museum in London as a historic instrument! The radial-velocity mask is in a long horizontal cell half-way up the picture; below it is a large $f/1$ aspheric Fabry lens (readily made by turning it from Perspex (methyl methacrylate) on a lathe) which images the grating on the central one of three photomultipliers in cylindrical housings below. The other photocells receive comparison regions of the spectrum; their signals are used to normalize the radial-velocity signal as the light transmission at the entrance slit fluctuates because of seing etc during the slow scanning of the mask.

The whole system of mask and photomultipliers is kinematically mounted with one degree of freedom remaining, and is scanned in the dispersion direction by the micrometer screw in the unit to the left of the main box. A 'Meccano' chain (from a children's constructional toy) is looped round three screws which provide focus travel; another runs in a figure-of-eight round the nuts of two of the screws to allow for tilt adjustment of the focus. In use, the photomultipliers are enclosed in a box that is cooled by continuous exchange of air with a refrigerating system behind him; the whole is then enclosed in a counterweighted cover that is raised from below on guide rails and is made of μ-metal to exclude the Earth's magnetic field as well as keeping the instrument light-tight.

Evidences for interaction among wide binary systems: To Ba or not to Ba?

Alain Jorissen *   *Henri M.J. Boffin* [†]

Abstract

The question as to whether binarity is a necessary and sufficient condition to produce a barium star is addressed. Since southern barium stars with strong anomalies are all binaries, binarity is truly required to form a barium star. Arguments supporting the wind accretion scenario as the origin of the barium syndrome are reviewed, and its predictions are discussed.

The presence of normal giants in binary systems with barium-like orbital parameters is a strong indication that binarity is not a sufficient condition to produce a barium star. It is suggested that barium stars form mainly in low-metallicity systems.

1. Introduction

Since the discovery by McClure et al. (1980) that barium stars belong to binary systems, this class of peculiar G and K giants exhibiting strong lines from heavy elements (e.g. Sr, Y, Zr, Ba, La, Ce) proved to be a unique laboratory for studying binary star evolution in wide spectroscopic systems. A paper devoted to barium stars has clearly its place in a conference honouring R.F. Griffin, who published in 1981 the first orbital elements for a barium star, suggesting at the same time the possible importance of binarity in the barium syndrome. Much progress has been done since, and it is now possible to answer the fundamental question as to whether binarity really is a *necessary and sufficient* condition to produce a barium star. In other words, (i) are all barium stars binaries? and (ii) do all stars belonging to binary systems with barium-like characteristics exhibit the barium syndrome? It will be shown that binarity is indeed a necessary condition for producing G and K giants with strong barium anomalies. Actually, the presence of chemical peculiarities at the

---

*European Southern Observatory, K. Schwarzschild Straße 2, D-W-8046 Garching bei München, Germany.

[†]Institut d'Astronomie et d'Astrophysique, Université Libre de Bruxelles, Av. F.D. Roosevelt 50, B-1050 Bruxelles, Belgium.

surface of these giants led to the recognition that some kind of interaction can occur between the components of rather wide binary systems. Wind accretion is one such possible interaction, whose properties will be reviewed in Sect. 3. Incidentally, some alteration of the primordial orbital elements occurs even for those rather wide binary systems.

Evidences against binarity as a sufficient condition for producing a barium star will then be presented in Sect. 4. The systems ξ^1 Ceti (HD 13611) and o Tau (HD 21120) are probably the best examples of evolved binaries where the former occurrence of wind accretion is not emphasized by chemical peculiarities. In the framework of the present conference centered on star formation, this conclusion is important in the sense that these unconspicuous binaries with evolved orbital elements may complicate the task of assembling a clean sample of unevolved binaries with the purpose of deriving constraints on star formation mechanisms.

2. Binarity: a required property for barium stars

Barium stars were defined as a class by Bidelman & Keenan (1951). At low dispersion, these peculiar giants (representing about 1% of all G – K giants; MacConnell et al. 1972) exhibit enhanced features of BaII, SrII, CH, CN, and sometimes C_2 (see McClure 1984a for a review of their spectroscopic and photometric properties). The BaII $\lambda4554$ line strength is generally used to characterize the abundance peculiarity, on a scale ranging from 1 (*mild* barium stars) to 5 (*strong* barium stars) originally introduced by Warner (1965). The kinematic behaviour of barium stars is similar to that of A – F main sequence stars, and their average mass should accordingly lie in the range 1.5 to 2.0 $M_\odot$ (Hakkila 1989a). Moreover, there appears to be a correlation between the level of chemical peculiarity and the age of the population, in the sense that mild barium stars appear to belong to a population younger than strong barium stars (Catchpole et al. 1977, Kovács 1985, Lü 1991).

Absolute magnitude determinations for individual barium stars are difficult. In the Hertzsprung – Russell diagram, the few stars with a reliable absolute magnitude are located either along the first red giant branch or in the core He-burning clump (Scalo 1976, Lambert 1985, Hakkila 1990).

The overabundances of trans-iron elements observed in barium stars clearly bear the signature of the s-process of nucleosynthesis (e.g. Cowley & Downs 1980, Tomkin & Lambert 1983, Smith 1984, Tomkin & Lambert 1986). In this process, heavy elements are synthesized by successive neutron captures starting from the iron-peak nuclei present in the star since its birth (see Käppeler et al. 1989 for a review). The necessary neutrons can be liberated in He-burning environments, either by the $^{14}N(\alpha, \gamma)^{18}F(\beta^+)^{18}O(\alpha, \gamma)^{22}Ne(\alpha, n)^{25}Mg$ chain if $T \geq 3\ 10^8$ K or by the $^{12}C(p, \gamma)^{13}N(\beta^+)^{13}C(\alpha, n)^{16}O$ chain if protons can somehow be mixed within the He-layer. As far as barium stars are concerned, the helium core flash is the only former evolutionary stage where neutrons might possibly have been liberated by the latter mechanism (see the discussion by Jorissen & Arnould 1989). Besides the fact that the operation of the s-process during the helium core flash is far from being well established, this explanation for the heavy element overabundances of barium stars also faces the difficulty that not all barium stars might have gone yet through the helium core flash, as suggested by their location in the HR diagram. The excesses of heavy elements present at the surface of barium stars thus remained a puzzle in the framework of single-star evolution.

The discovery that barium stars belong to binary systems (McClure et al. 1980, Griffin & Griffin 1981, McClure 1983, Jorissen & Mayor 1988, McClure & Woodsworth 1990, Griffin 1991) opened a wide range of possibilities for explaining the origin of the barium-star chemical peculiarities. The fact that binarity is a necessary condition to produce a barium star has been assessed by a CORAVEL monitoring (Jorissen & Mayor, 1988 and in preparation) that revealed the binary nature of all 27 barium stars with strong anomalies (Ba4 or Ba5) known in the southern hemisphere (Lü et al. 1983). Since periods as long as 10 y are observed among barium systems (corresponding to an orbital separation of the order of 1000 $R_\odot$, assuming a total mass of 2 $M_\odot$), it seems unlikely that the companion could have had a direct influence on the internal structure of the barium star. Therefore, one may consider instead scenarios involving the pollution of the barium-star envelope by nuclearly processed matter from its companion. The nature of the unseen companion then plays a crucial role for assessing the validity of the mass transfer hypothesis.

No information on the companion can be obtained from the visible part of the spectrum where only red giant features are apparent. Among the 11 barium stars surveyed in the far UV (excluding HD 65699 and ξ^1 Ceti, which are no more considered as barium stars; see Smith & Lambert 1987a and Sect. 4), excess light due to a white dwarf (WD) companion was detected for ζ Cap and possibly for ζ Cyg (Böhm-Vitense 1980, Dominy & Lambert 1983, Böhm-Vitense et al. 1984), while the origin of the excess UV light arising from the 56 Peg system is still debated (WD or accretion disk? Schindler et al. 1982, Dominy & Lambert 1983). Fekel (in Strassmeier et al. 1988) also reported the presence of a hot companion around the mild barium star HD 165141 (K0IIIBa1). But it is undoubtedly the analysis of the mass-function distribution that has proven to be the most useful way to date to gain knowledge about the companion. Webbink (1986), McClure & Woodsworth (1990) and Boffin et al. (1992) showed that the mass-function distribution of barium stars can be fitted quite well by a sample of orbits with $Q = M_2^3/(M_1 + M_2)^2 = 0.04 \pm 0.01$ $M_\odot$ seen at random inclinations (M_1 and M_2 are the masses of the barium star and of its companion, respectively). Such a sharp Q-distribution is consistent with companions being WD's with masses spanning a narrow range, as is the case for field WD's. If WD companions have masses in the range 0.55 ± 0.05 $M_\odot$ (e.g. Bergeron et al. 1991), then masses of 1 to 2 $M_\odot$ for the barium star would yield Q values in the observed range (see Jorissen & Mayor 1992). The absence of far-UV excess for most of the barium stars then indicates that their WD companions are rather cold ($T_{\text{eff}} < 10^4$ K). Since the cooling timescale of such WD's exceeds the lifetime of a low-mass star in the giant phase, Johnson et al. (1992) concluded that barium stars were formed while still on the main sequence. This important prediction will be discussed further in Sect. 3.2.2.

The WD nature of the companion provides an important clue to the former history of the system, since it means that sometime in the past, an asymptotic giant branch (AGB) star was present in the system. Asymptotic giant branch stars, like carbon stars or Tc-rich S stars, are known to lose mass through a strong wind (e.g. Knapp 1985, Jura 1988, Wannier et al. 1990), and to be heavy element-rich (Utsumi 1985, Smith & Lambert 1990). The heavy element Tc, which has no stable isotopes, is often observed at the surface of S and C stars (Little et al. 1987, Smith & Lambert 1988). This observation is a clear indication that heavy-element synthesis is presently going on in these AGB stars, because the isotope ^{99}Tc involved in the s-process has a half-life of only $2 \ 10^5$ y at the stellar surface, which is of the same order as the AGB lifetime.

Transfer of heavy element-rich matter from the AGB companion towards the barium star thus appears to be an attractive explanation for the chemical peculiarities of the latter. Mass transfer is possible either through Roche lobe overflow (RLOF) or wind accretion. Before presenting in Sect. 3 evidences supporting the latter mechanism, it should be noted that a similar mass transfer scenario can be invoked for other families of peculiar red giants that were also found to belong to binary stars. These are the CH giants, belonging to Pop II (McClure 1984b, McClure & Woodsworth 1990), the CH subgiants, which are less evolved than barium stars on the first giant branch (Bond 1984, McClure 1985, 1989, Luck & Bond 1991), and S stars without Tc (Smith & Lambert 1987b, Jorissen & Mayor 1988, 1992, Brown et al. 1990, Johnson 1992). These S stars without Tc, also referred to as extrinsic S stars (whereas the genuine Tc-rich S stars populating the AGB are sometimes called intrinsic S stars), have WD companions like barium stars (Johnson et al. 1992, Jorissen & Mayor 1992). Therefore, they are probably the cooler counterparts of barium stars, populating the tip of the red giant branch.

3. The wind accretion scenario

3.1. Case for support

3.1.1. Comparison of abundances

If the former AGB companion is responsible for the pollution of the barium star envelope, the chemical peculiarities of the latter should then resemble those of the

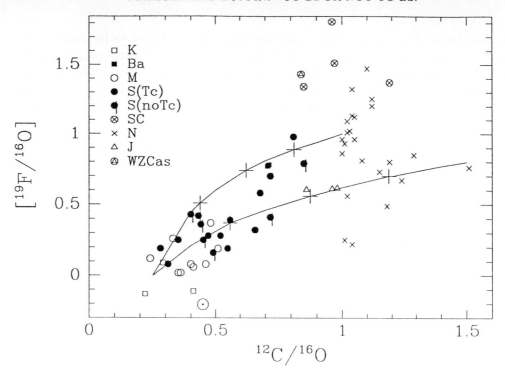

Figure 1.   Correlation between fluorine abundances and $^{12}C/^{16}O$ ratios in several families of late-type giants (Jorissen et al. 1992b). The solid lines represent the evolution of the envelope composition of barium stars (or Tc-deficient S stars) as an increasing amount of material from a N-type carbon star (upper-right end of the curve) is added to the envelope of a normal K giant (lower-left end of the curve). The crosses on the mixing lines refer to dilution factors $r = 0.25$, 0.50, and 0.75, respectively (see Eq. 1).

former. These chemical peculiarities will be assumed to be similar to those of present-day AGB stars with metallicities $-0.8 \leq [Fe/H] \leq 0$ like in barium stars.

Fluorine has recently been identified as a sensitive probe of AGB nucleosynthesis (Jorissen et al. 1992b, Forestini et al. 1992), since its overabundance is tightly correlated to the increase of $^{12}C/^{16}O$ observed along the AGB. Fluorine and carbon are thus produced together in the He-burning layer of AGB stars. Unlike K giants where F has a normal (i.e. solar) abundance, barium stars exhibit large F overabundances (Fig. 1), thus suggesting that barium stars were contaminated by matter from carbon stars. The same holds true for extrinsic S stars, since they cannot be distinguished from intrinsic S stars on the basis of their position in Fig. 1. Similarly, Lambert (1988) showed that the heavy-element and carbon abundances of barium stars can be reproduced by diluting carbon-star matter in a normal K-giant envelope. Figure 1 can be used to estimate the amount M_{acc} of matter accreted by the barium star.

Assuming that the accreted matter has been completely mixed in the barium-star convective envelope of mass M_{env}, the overabundance g_i of species i observed at the surface of barium stars relates to the overabundance f_i in the accreted matter through the expression

$$g_i = \frac{M_{env} + f_i M_{acc}}{M_{env} + M_{acc}} = (1 - r) + r f_i, \qquad (1)$$

with $r = M_{acc}/(M_{acc} + M_{env})$ being the dilution factor and with the overabundances f_i and g_i being expressed with respect to the normal K-giant envelope before contamination. The dilution factors deduced from Fig. 1 or from Lambert (1988) are of the order of 0.5 for strong barium stars and < 0.2 for mild barium stars. In other words, if the present envelope mass amounts to ~ 1 $M_{\odot}$ (Sect. 2), a few 0.1 $M_{\odot}$ up to 1 $M_{\odot}$ in the most extreme cases had to be accreted in order to produce the observed overabundances.

One should keep in mind, however, that the comparison between the barium- and carbon-star compositions might be blurred by several effects, like chemical segregation between volatile and refractory elements if gas and grains were to be accreted with different efficiencies. Moreover, matter has been accreted by the barium star during the whole mass-losing episode of its AGB companion, which is not restricted to the carbon-star phase. The composition of the accreted matter is thus an average over the surface compositions experienced by the companion during its whole AGB lifetime. On the other hand, the mixing of the accreted matter within the barium-star envelope might have involved deep layers partially processed by the CN cycle (Proffitt 1989, Proffitt & Michaud 1989), so that Eq. 1 would not be valid for species like ^{13}C or ^{14}N for example. Barbuy et al. (1992) have argued that such an effect could perhaps explain why some barium stars have $^{12}C/^{13}C$ ratios lower than expected from the accretion of ^{12}C-rich matter from a carbon star. Finally, although Tc is present in AGB stars, its absence in barium stars (Little-Marenin & Little 1987) is compatible with the long time span between the present state of the system and the mass transfer event, as deduced from the WD cooling timescale (Sect. 2).

3.1.2.   Dynamics of mass transfer

The wind accretion scenario initially appeared as an interesting alternative to mass transfer through RLOF that suffers from two severe difficulties in the context of barium star evolution.

First, there is empirical evidence that the orbits of post-RLOF systems are always circular (e.g. Popper 1989 for Algol systems, Burki & Mayor, 1983 and **in preparation** for yellow supergiants), and this situation contrasts with that prevailing for barium systems.

Figure 2 compares the $(e, \log P)$ diagrams of barium stars and open cluster giants,

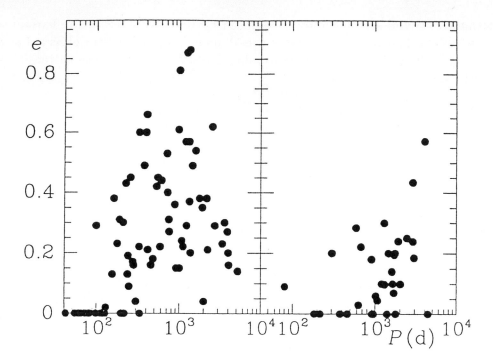

Figure 2.   Diagram (e, log P) for barium stars (right panel, from McClure & Woodsworth 1990 and Jorissen & Mayor 1988, 1992 and in preparation) and late-type giants in clusters (left panel, from Mermilliod & Mayor, this conference).

and clearly shows that most barium systems have non-circular orbits, though with an average eccentricity smaller than that of normal cluster giants at any given period. In the case of post-RLOF systems, the orbital circularization has likely been achieved through tidal effects operating when the mass-losing star was about to fill its Roche lobe, since the circularization timescale is then very short with respect to the evolutionary timescale (Zahn 1966, 1975, 1977, 1989, Lecar et al. 1976, Tassoul 1988, Rieutord, this conference).

The second difficulty with the RLOF scenario is related to the fact that the process is dynamically unstable when the mass loser is an AGB star with a deep convective envelope and with a masss exceeding 0.7 times that of its companion (Paczyński 1976, Tout & Hall 1991). A main sequence star cannot stand the very large mass transfer rate resulting from that process without expanding to form a contact system[1] (e.g. Meyer & Meyer-Hofmeister 1979). Such a common envelope evolution leads to

---

[1] Note that such a difficulty would not exist for accreting giant stars, which can stand much larger accretion rates than dwarfs due to their shorter Kelvin-Helmoltz timescale. However, arguments based on the WD cooling timescale (Sect. 2) indicate that the accretion event responsible for the barium syndrome must have occurred when the proto-barium star was still on the main sequence.

degenerate systems like cataclysmic variables. Since the barium system did avoid this dramatic fate, it ought to remain detached when the companion was on the AGB. This argument thus calls for pollution of the barium star through wind accretion rather than through RLOF. However, given the present periods of barium systems (Fig. 2), it is not immediately apparent that this key requirement could be fulfilled. Due to the large amount of mass lost by the WD progenitor, a typical barium system with $P = 1000$ d, $M_1 = 1.5$ M$_\odot$ and $M_2 = 0.6$ M$_\odot$ (corresponding to $A = 540$ R$_\odot$) was much closer in the past. Assuming a Jeans' mass loss mode (corresponding to $A(M_1 + M_2) = $ cst), the orbital separation was only 320 R$_\odot$ when $M_2(0) = 2$ M$_\odot$, so that the Roche radius $R_{R,2} = A(0.38 + 0.2 \log(M_2(0)/M_1))$ (Paczyński 1971) amounted to 130 R$_\odot$. In comparison, a typical AGB star with $\log T_{\text{eff}} = 3.5$ and $M_{\text{bol}} = -5$ has a radius of 280 R$_\odot$! This difficulty can be circumvented if the wind mass loss rate of the giant is enhanced by the presence of the companion, as suggested by Tout & Eggleton (1988). In such conditions, the giant loses its envelope before filling its Roche lobe, and these authors show that systems with *final* periods as short as 100 d remained always detached (for a primary of initial mass 2 M$_\odot$). It is interesting to note that all barium systems precisely lie above this threshold period (Fig. 2), although systems with shorter periods and consisting in a giant and a degenerate companion nevertheless exist (like HD 185510 with $P = 20.7$ d, HD 160538 with $P = 31.5$ d and AY Cet with $P = 56.8$ d; Fekel & Simon 1985, Simon et al. 1985). High-resolution spectroscopy of the systems HD 185510 and AY Cet (Jorissen, unpublished) using the method described by McWilliam (1990; see also Sect. 4) showed that they are not heavy element-enriched like barium stars. The lower period cutoff observed among barium stars must therefore correspond to a real transition between different evolutionary schemes.

The previous discussion, by emphasizing the difficulties encountered by RLOF scenarios, provided support to the wind accretion scenario *ad absurdum*. It still remains to prove that the wind accretion process can produce barium stars with the required properties. A key ingredient of the wind accretion scenario is obviously the mass accretion rate. Previous applications of that scenario (Livio & Warner 1984, Boffin & Jorissen 1988) made use of the Bondi–Hoyle accretion rate (Bondi & Hoyle 1944). In this scheme, the gravitational potential of the accreting star transforms a supersonic plane-parallel flow at infinity into symmetrically colliding streams along the leeward axis. Matter is accreted from the accretion column that forms on this axis, at a rate

$$\dot{M} = \alpha \pi R_a^2 \rho_\infty v_\infty \qquad (2)$$

where ρ_∞ and v_∞ are the density and relative velocity of the flow at infinity, respectively, $R_a = 2GM/v_\infty^2$ is the accretion radius, G is the constant of gravitation, M is the mass of the accreting star, and α is a numerical constant between 0.5 and 1 whose value cannot be fixed by the theory. This formula can be adapted to the context of binary systems to relate the amount of accreted mass dM_2 to the amount of mass

Table 1. Wind accretion rate $-dM_2/dM_1$ predicted by the Bondi–Hoyle theory, if $\alpha = 0.5$, $M_1 = 3$ $M_\odot$ and $M_2 = 1.5$ $M_\odot$. The numbers in parenthesis give R_a/A, the fractional accretion radius.

P	A	v_{orb}	v_w (km s^{-1})	
(y)	(AU)	(km s^{-1})	15	30
2.4	3.	36	0.11 (0.57)	0.03 (0.40)
6.9	6.	26	0.06 (0.50)	0.01 (0.29)

dM_1 lost by the primary star, as follows (Livio & Warner 1984)

$$dM_2/dM_1 = -\frac{\alpha}{(1 - e^2)^{1/2} A^2} \left(\frac{GM_2}{v_w^2}\right)^2 \frac{1}{[1 + (v_{orb}/v_w)^2]^{3/2}}, \qquad (3)$$

where v_w and v_{orb} are the wind and orbital velocities, respectively, both expressed relative to the mass-losing star. A blind application of this formula to the conditions prevailing in the proto-barium systems leads to rather large accretion rates, as shown in Table 1, which certainly make this scenario promising.

It is important to note, however, that the assumptions inherent to Bondi–Hoyle's accretion rate are not necessarily fulfilled in the proto-barium systems, mainly because $v_w \sim v_{orb}$. In such conditions, wind particles follow complicated three-body trajectories, and the existence of colliding streams as in the Bondi–Hoyle picture must be questioned. The accretion radius R_a (Eq. 2), of the order of the orbital separation (Table 1), further illustrates the problem.

A definite demonstration of the merits of the wind accretion scenario therefore requires the use of more sophisticated numerical methods, like smooth particle hydrodynamics (SPH; Lucy 1977, Gingold & Monaghan 1977). Such 3D-calculations are presently in progress, and yield an accretion efficiency $-dM_2/dM_1 = 0.04$ in the case $A = 3$ AU and $v_w = 15$ km s^{-1} (adiabatic gas), to be compared with 0.11 reported in Table 1 for Bondi–Hoyle's accretion rate. More details will be reported elsewhere (Theuns et al., in preparation). For a 3 $M_\odot$ primary losing about 2.5 $M_\odot$, the SPH rate thus predicts that about 0.1 $M_\odot$ can be accreted by the barium star. As indicated in Sect. 3.1.1, this value is just large enough to account for the level of chemical peculiarities exhibited by mild barium stars. Larger accretion rates are required for strong barium stars, but as discussed in Sect. 3.2.1., they generally belong to systems with shorter periods than the one that was modelled, and larger accretion rates may accordingly be expected. A more detailed discussion of the merits of the wind accretion scenario, based on these new SPH rates and evaluating the eccentricity variations, is deferred to a forthcoming paper.

3.2.   Predictions of the wind accretion scenario

3.2.1.   Do mild barium stars belong to wider systems than strong barium stars?

In the framework of the wind accretion scenario, the accretion rate obviously depends upon the orbital separation (Eq. 2 and Table 1). Therefore, barium stars belonging to wide systems may obviously be expected to have been contaminated by the wind of their companion to a lesser extent than those in closer systems. It is therefore important to look for a possible correlation between the level of chemical anomalies and the orbital period, both quantities being directly accessible to the observation.

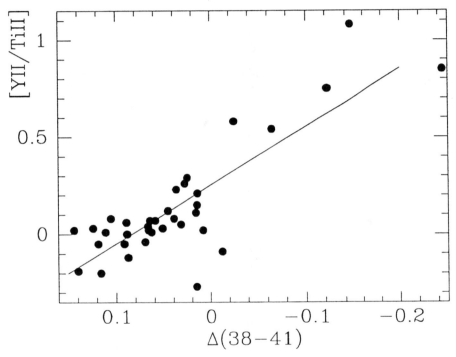

Figure 3.   The $\Delta(38 - 41)$ color index vs. [YII/TiII] for G – K giants with DDO colors from Lü (1991) and abundances from McWilliam (1990). The solid line results from a least square fit to the data.

The level of chemical peculiarities in a barium star is usually expressed in terms of the barium index, originally introduced by Warner (1965), and ranging from Ba1 to Ba5 with increasing peculiarities. Since it relies on a visual estimate of the BaII λ 4554 line strength, it is not a very objective quantity to rely on. A photometric index constructed from DDO colors will therefore be used instead of the barium index. Lü et al. (1983) and Lü (1991) showed that there is a strong correlation between the position of a barium star in the $C(38 - 41)$ vs. $C(42 - 45)$ plane and its barium index (where $C(i - j)$ is the color index corresponding to filters i and j of the DDO system; see McClure & van den Bergh 1968). More precisely, an index of the form $\Delta(38 - 41) = C(38 - 41) - m\, C(42 - 45) + n$ should be related to the level of chemical peculiarities (see Fig. 4 of Lü 1991). That index can actually be calibrated in terms of [YII/TiII] abundances, using the 35 G and K giants for which both DDO colors (Lü 1991) and abundances (McWilliam 1990) are available. McWilliam (1988, 1990) showed that the abundance ratio YII/TiII, derived from the YII $\lambda 6795.4$ and TiII $\lambda 6607.0$ lines, is a very good indicator of heavy element abundances in red giants, since it is relatively insensitive to the atmospheric parameters. The distribution of YII/TiII ratios in a sample of about 600 G – K giants is very narrow, centering at $\log(\text{YII/TiII}) = -2.75$ (adopted as normalization for [YII/TiII]) with a tail extending only 0.25 dex upwards. The relation between [YII/TiII] and $\Delta(38 - 41)$ has therefore been normalized in such a way that barium stars ([YII/TiII] > 0.25) have $\Delta(38-41) <$ 0. The values $m = 0.89$ and $n = 1.32$ (yielding [YII/TiII] $= -3.02\Delta(38 - 41) + 0.25$; Fig. 3) result from a least-square fit between [YII/TiII] and $\Delta(38 - 41)$ for the 35 giants in common between the samples of Lü (1991) and McWilliam (1990). Mild barium stars then typically have $-0.1 \le \Delta(38 - 41) \le 0$ while strong barium stars (with Ba4 and Ba5 indexes) have $\Delta(38 - 41) \le -0.1$.

It has to be noted that $\Delta(38 - 41)$ is not very sensitive to interstellar reddening, since $\Delta^\star(38 - 41) = \Delta(38 - 41) - 0.08 E_{B-V}$, where $\Delta^\star(38 - 41)$ is constructed from the de-reddened indices as defined by Lü & Upgren (1979). The $\Delta(38 - 41)$ index defined from DDO colors is closely related to the Δc_1 index from Strömgren photometry described in Jorissen et al. (1992a).

Figure 4 shows that only strong barium stars are found among short period systems (Note that HD 77247, G5Ba1 with $P = 80$ d, could not be plotted, because DDO colors are not available). There is nevertheless a large scatter in the relation between orbital period and chemical peculiarity. Several factors may explain that scatter, namely (i) $\Delta(38 - 41)$ may be sligthly dependent upon parameters other than the chemical composition, like effective temperature, luminosity or metallicity; (ii) the orbital period is related to the orbital separation (which is more directly connected to the efficiency of the wind accretion process) through the mass of the system, which may obviously vary from one system to the other; (iii) the mass of the barium-star envelope, in fixing the dilution factor of the accreted matter, is another key parameter, and finally (iv) not all former AGB companions were equally efficient in synthesizing (or dredging-up in their envelope) heavy elements.

The comparison between HD 121447 ($P = 185$ d, $\Delta(38 - 41) = -0.265$, K7Ba5) and HD 178717 ($P = 2866$ d, $\Delta(38 - 41) = -0.200$, K4Ba5) is illustrative in that respect. Despite extremely different orbital periods, their level of chemical peculiarities is very similar, as shown by their $\Delta(38 - 41)$ index and their fluorine (Fig. 1) or heavy element contents (Smith 1984). Assuming moreover similar masses for both stars (as

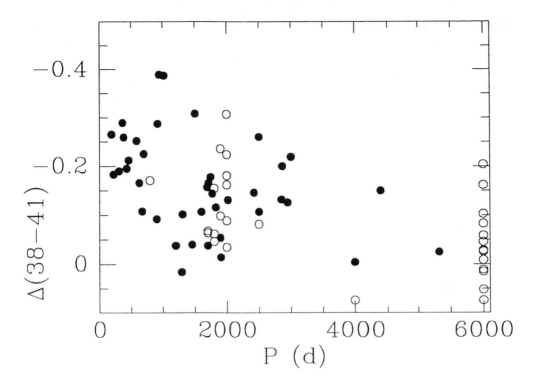

Figure 4. Orbital period vs. $\Delta(38 - 41)$, for the combined sample of barium stars from McClure & Woodsworth (1990) and Jorissen & Mayor (1988 and in preparation). In this conference in honour of R.F. Griffin, one should emphasize that the longest period known among barium stars (16 Ser = HD 139195; $P = 5324$ d) results from his Cambridge measurements (Griffin 1991). Open circles correspond to systems for which only a lower limit is available for the orbital period. A period of 6000 d has been arbitrarily assigned to systems with no radial velocity variations.

suggested by similar luminosities, gravities and effective temperatures; Eggen 1972, Smith 1984), this difference in periods actually translates into a difference in orbital separations. In order to achieve similar levels of chemical peculiarities despite the much smaller accretion rate of the wider system, the matter accreted by HD 178717 must have been much more heavy element-rich than that accreted by HD 121447. The difference in metallicity between HD 121447 ([Fe/H] = +0.05) and HD 178717 ([Fe/H] = -0.18) obtained by Smith (1984) may be the clue to the different efficiencies of heavy-element synthesis in the former AGB companion of these stars. As will be discussed further in Sect. 4, there are indeed some indications that heavy-element synthesis (or dredge-up in the envelope) is more efficient in low-metallicity AGB stars.

3.2.2.   Dwarf barium stars

A key prediction of the wind accretion scenario is the existence of dwarf barium stars. Since the efficiency of wind accretion does not depend on the star radius (Eq. 2), accretion by dwarfs is in principle as efficient as accretion by giants (neglecting magnetic effects that may be important in dwarfs but not in giants). A crucial difference between dwarf and giant accretors, already emphasized in Sect. 3.1.2, arises from their different Helmoltz-Kelvin timescales $\tau_{HK} = 2 \; 10^7 M^2/LR$ (where the luminosity L, mass M and radius R of the accreting star are expressed in solar units, and τ_{HK} is in y), since rapid mass accretion (i.e. $M/\dot{M} < \tau_{HK}$, or $\dot{M} \geq 5 \; 10^{-8} LR/M$) leads to the expansion of the matter dumped on the radiative envelope of a main sequence star (e.g. Kippenhahn & Meyer-Hofmeister 1977, Neo et al. 1977). According to the former authors, accretion rates of the order of 10^{-4} M$_\odot$ y^{-1} (on a 2 M$_\odot$ main sequence star) are nevertheless required to induce the large radius increase leading to contact systems. Therefore, wind accretion rates of the order of $0.1\dot{M}_{AGB}$ (Sect. 3.1.2), with $\dot{M}_{AGB}$ ranging from 10^{-8} M$_\odot$ y^{-1} for optical S and C stars to 10^{-4} M$_\odot$ y^{-1} for OH/IR stars (Knapp & Morris 1985, Wannier et al. 1990), should allow the main sequence accretors to remain detached in the proto-barium systems.

Several dwarf barium stars are actually known, as summarized in Table 2. Other candidates, not listed in Table 2, can be found among the F dwarfs with strong Sr listed by Cowley (1976) as well as in Houk's *Michigan Catalogue of Two-Dimensional Spectral Types for the HD stars* (North, private communication). Although some of the stars listed in Table 2 do not show any evidence for binarity (thus constituting a challenge[2] for future studies), good candidates for the long-sought dwarf progenitors of barium stars can be found among the F strong λ 4077 stars (North & Duquennoy, 1991 and this conference), which are peculiar F dwarfs first identified by Bidelman (1981, 1983, 1985), as well as among the CH subgiants, some of them being actually dwarfs (Luck & Bond 1991). The frequency of str. λ 4077 stars among F dwarfs is similar to that of barium stars among G – K giants, but a definite proof of the evolutionary link between these families should await the availability of orbital elements.

A search for WD companions around 3 F str λ 4077 stars (North & Lanz 1991) and 21 CH subgiants (Bond 1984) yielded negative results. North & Lanz show that their result can still be accomodated with the timescale constraints (which are now more severe than in the case of giant barium stars; see Sect. 2), especially so if the considered dwarfs are old, as indicated by their slight metal-deficiency. This argument will be developed further in Sect. 4, by presenting several evidences that barium stars do not form in young, metal-rich systems. The failure to detect any WD's around 21 subgiant CH stars is more puzzling, however, and could actually endanger the mass transfer scenario. Smith & Lambert (1986) suggested that this failure must perhaps

---

[2]See, however, Lambert 1985 who questions the reality of the heavy-element overabundances in ξ Boo A and ϵ Indi

Table 2. Dwarf barium (or carbon) stars

(i) in binary systems

name	Ref.	Sp.	[Ba/Fe]	P (d)	e	$f(M)(M_\odot)$	Ref
F str. $\lambda4077$	ND91	FV	–	several SB			ND91
sgCH	LB91	F,GV	~ 0.7	several SB			LB91
HR 4395	T90	F5	0.8	1940 ± 84	0.6 ± 0.5	0.03	AL76,MG87
G77-61	D77	CV	0.0?	245	< 0.2	$0.3 + 0.55$	D86

(ii) with no evidence for binarity

| name | Ref. | Sp. | [Ba/Fe] | radial velocity | | | | Ref |
				σ (km/s)	E/I	N	Δt (d)	
HR 107	T89	F6V	0.4	0.52	1.5	13	2666	COR, ND91
ζ Boo A	BE84	G8V	0.2	0.21	0.5	14	4414	COR
HD 130255	LCS91	K0 (sgCH)	–	0.43	1.1:	29	3305	McW90
ϵ Indi	K80	K4-5V	0.1	0.06	0.3	3	414	COR
HD 6434	T89	G2V (sgCH)	0.4					
LP 328-57	G91	CV						
LP 756-18	G91	CV						
CLS 31	G91	CV						
	L91	F-KV						

References:
AL76: Abt & Levy 1976; BE84: Boyarchuk & Eglitis 1984; COR: CORAVEL database (unpublished); D77: Dahn et al. 1977; D86: Dearborn et al. 1986; G91: Green et al. 1991; K80: Kollatschny 1980; L91: Lü 1991; LB91: Luck & Bond 1991; LCS91: Lambert et al. 1991; McW90: McClure & Woodsworth 1990; MG87: Morbey & Griffin 1987; ND91: North & Duquennoy 1991; T89: Tomkin et al. 1989; T90: Tomkin, 1990, private communication

Radial velocity data:
σ is the standard deviation of the radial velocity measurements, $E/I = \sigma/\bar{\epsilon}_1$ (where $\bar{\epsilon}_1$ is the average uncertainty on one measurement) is a good indicator of variability (see Jorissen & Mayor 1988), N is the number of measurements and Δt the time span of the measurements.

be attributed to dust surrounding the WD. Infrared excesses indeed point towards the presence of dust in the barium systems (Bartkevicius & Sviderskiene 1981, Hakkila & McNamara 1987, Hakkila 1989b), and recent observations of long-term photometric variations for the barium star HD 46407 (Jorissen et al. 1991, 1992a) may indicate that this dust actually surrounds the companion. The observed variations are roughly in phase with the orbital period ($P = 458$ d; McClure & Woodsworth 1990). The lightcurve exhibits a wide dip of depth $\Delta y \sim 0.01$ mag centered on the time of transit of the companion in front of the barium star, but extending over about $\sim 40\%$ of the orbital period. Since the barium star then also appears redder ($\Delta(v-b) \sim 0.01$ mag), scattering on dust grains surrounding the companion could possibly account for the observed variations. The extinction at 1600 Å derived from the above mentioned reddening, using the relations provided by Nandy et al. (1976, for interstellar dust) and Crawford & Mandwedala (1976), is of the order of 0.1 to 0.2 mag, a value too small to account for the non-detection of the UV light from a hot WD companion. Nevertheless, Böhm-Vitense & Johnson (1985) suspected some absorption of the UV light to be present in the ξ^1 Cet system (see Sect. 4), because the observed UV flux is smaller than that predicted for a WD having the effective temperature and gravity derived from the shape of the UV continuum.

4.   Binarity is not sufficient to produce a barium star

Table 3a presents binary systems with a red giant primary and with orbital parameters similar to those of barium stars, extracted from the list[3] of Boffin et al. (1992). All systems in Table 3a meet constraints characterizing barium-like systems, namely (i) $200 < P(\mathrm{d}) < 6000$, (ii) $f(M) \leq 0.1$ M$_\odot$ and (iii) $e < e_{\mathrm{Ba}}(P)$, where $e_{\mathrm{Ba}}(P)$ is the maximum eccentricity observed at period P among barium systems (Fig. 2). Constraint (i) ensures that the considered systems followed the same binary evolution as barium stars. Systems consisting in a red giant and a degenerate component like HD 185510 (K0III-IV + sdB; $P = 20.6$ d), HD 160538 (K0-2III + WD; $P = 31.5$ d) and AY Ceti (G6IIIe + WD; $P = 56.8$ d) mentioned by Eggleton (1986) and Fekel & Simon (1985) in a similar context, were thus not retained because of their short orbital periods. Constraint (ii) allows to get rid of systems with massive main sequence companions, while systems not fulfilling constraint (iii) are likely unevolved (see Fig. 2 and Boffin et al., this conference) and therefore contain a main sequence rather than a WD companion. On the other hand, it is very likely that binaries with $e < 0.15$ and $P > 100$ d contain a WD companion, since samples of unevolved systems never populate that region of the $(e, \log P)$ plane (see Fig. 2 and **Mermilliod & Mayor**, this conference, for cluster giants, and Duquennoy & Mayor 1991 for systems with G dwarf primaries). Constraints (i) and (ii) do not allow, however, to totally rule out systems with low-mass main sequence companions in the list of Table 3a. IUE spectra provide direct evidences for a WD companion only in the case of HD 13611

---

[3]Except for HD 81817, but CORAVEL measurements are in progress

Table 3a. Normal red giants in barium-like binary systems

HD	HR	Sp. Type	P (d)	e	f(m) (M$_\odot$)	Ref.	[Fe/H]	[YII/TiII]	[LaII/TiII]
11559	549	K0 III	1672.4	.18	.0165	3	-0.11	0.15	0.11
13520	643	K3.5 III	748.2	.34	.0075	3	-0.24	0.01	0.00
13611	649	G8 III + wd	1642.1	0.	.0035	3	-0.29	0.04	0.37
21120	1030	G8 III + wd ?	1654.9	.26	.0132	1	-0.15	0.26	0.38
21754	1066	K0 II-III	960.	.4	.0439	1	-0.12	0.22	0.46
27697	1373	K0 IIICN.5	529.8	.42	.0011	1	0.00	0.10	0.04
33856	1698	K0.5 III	1031.4	.1	.0695	1	0.06	0.00	0.12
49293	2506	K0 III	1760.9	.4	.009	1	-0.12	0.20	0.37
62721	3003	K4 III	1519.7	.325	.05	1	-0.27	0.05	-0.19
66216	3149	K2 III	2438.	.06	.035	1	0.03	0.01	-0.17
81817	3751	K3II + wd				2	0.09	0.00	-0.07
88284	3994	K0 IIICN1	1585.8	.14	.0081	1	0.05	0.02	0.26
90537	4100	G9 IIIab	13800.	.66	.02	1	0.00	-0.12	0.12
106760	4668	K0.5 IIIb	1314.3	.426	.0283	1	-0.29	0.00	0.12
120539	5201	K4 III	944.	.41	.0001	1	-0.24	-0.11	0.02
169156	6884	K0 III	2373.79	.1	.0474	1	-0.17	0.11	0.17
175515	7135	G9-K0 III	2994.	.24	.029	1	-0.26	0.05	0.33
176411	7176	K1 IIICN.5	1270.6	.27	.0162	1	0.00	-0.01	0.16
176524	7180	K0 III	258.48	.21	.0054	1	-0.12	0.12	0.26
196574	7884	G8 III	205.2	0.	.0201	1	-0.13	0.19	0.33
197752	7939	K2 III	2506.	.38	.0212	1	-0.19	-0.07	0.02
						Average:	-0.12±0.13	0.07 ± 0.10	0.15 ± 0.19

Table 3b. Comparison sample of barium stars

HD	HR	Sp. Type	P (d)	e	f(m) (M$_\odot$)	Ref.	[Fe/H]	[YII/TiII]	[LaII/TiII]
116713	5058	K1Ba3					-0.25	0.85	0.98
139195	5802	K1Ba1	5324	.345	.0263	4	-0.17	0.58	0.50
202109	8115	G8Ba< 1	> 1700			5	-0.11	0.54	0.52
204075	8204	G5Ba2	2422	.25	.0042	6	-0.45	1.08	1.21
					< [Fe/H] >=		-0.25 ±0.14		

Refs. 1: Batten et al. (1989); 2: Reimers (1984); 3: Griffin, R.F. (1985); 4: Griffin (1991); 5: Jorissen & Mayor, in preparation; 6: McClure & Woodsworth (1990)

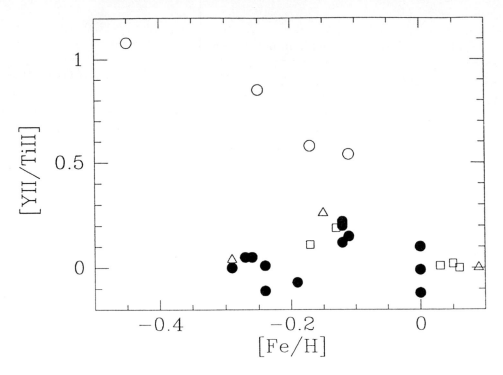

Figure 5. [YII/TiII] vs. metallicity [Fe/H] for the giants of Table 3a and 3b, with abundances taken from McWilliam (1990). Black dots refer to systems where the nature of the companion is not known, open squares stand for systems with $e < 0.15$ where the companion is suspected to be a WD, triangles refer to systems with known WD companions (HD 13611 = ξ^1 Ceti, HD 21120 and HD 81817) while open circles stand for barium stars.

(ξ^1 Ceti, Böhm-Vitense & Johnson 1985; see also the discussion below), HD 21120 (o Tau) and HD 81817 (Reimers 1984). The spectrum of HD 21120 (SWP 152121) has been retrieved from the IUE archives and is almost identical to that of ξ^1 Ceti.

Metallicities and heavy-element abundances have all been taken from McWilliam (1990), thus ensuring a good homogeneity of the abundance data listed in Table 3a. The [YII/TiII] and [LaII/TiII] ratios were normalized by the mean values of McWilliam's sample, namely log YII/TiII = −2.75 and log LaII/TiII = −4.0. For comparison, barium stars present in McWilliam's sample have been listed in Table 3b. Figure 5 shows that *none* among the systems of Table 3a exhibit Y overabundances as large as those of barium stars, although HD 21120, which is known to have a WD companion, lies at the edge of the [YII/TiII] values found in normal giants. But the other two systems with known WD companions, namely HD 81817 and HD 13611 (ξ^1 Ceti), are clearly not barium stars.

This then raises the question of the extra-property required to form a barium star, which would not be met by HD 13611 and HD 81817. Possibilities include the mass or the metallicity of the former AGB companion, because both properties could influence the ability of an AGB star to become heavy element-rich. An explanation in terms of metallicity appears likely for HD 21120, HD 81817 and for the systems with $e < 0.15$ (thus possibly having a WD companion as well; see above), because they all have a rather high metallicity ($[Fe/H] > -0.2$) when compared to barium stars (Fig. 5). Actually, there are several other arguments indicating that barium or carbon stars may be difficult to form in a high-metallicity population. Kovács (1985) already pointed out that there seems to be a correlation between barium overabundances and metallicities among barium stars themselves. The tendency of strong barium stars to be metal-poor seem to be reflected by McWilliam's data as well, as seen on Fig. 5. If real, this correlation would indicate that strong barium stars belong to an older population than mild barium stars. The kinematic analyses of Catchpole et al. (1977) and Lü (1991) support that conclusion, since these authors conclude that strong barium stars have larger velocity dispersions than mild barium stars. One might then speculate that AGB stars become more easily heavy element-rich if they are of low-metallicity and, therefore, barium stars would be more easily formed in a low-metallicity population. It is well known that the fraction of carbon stars (i.e. heavy element-rich stars) among AGB stars is higher in low-metallicity populations (e.g. Richer 1989, Jura 1991). This effect may simply result from the easiness to turn an oxygen-rich star into a carbon-rich star if the initial oxygen content of the envelope is low. If that were the case, this observation does not necessarily imply that heavy-element synthesis is more efficient in a low-metallicity population, and would thus not be very relevant to our purpose. However, the higher fraction of carbon stars in low-metallicity populations might also result from a direct effect of the metallicity on the stellar structure, favouring for example heavy-element synthesis or dredge-ups at low-metallicities.

Alternatively, the difficulty of forming barium systems in a high-metallicity population might be related to the faster AGB winds encountered in that case, as shown by Zuckerman & Dyck (1989) for carbon stars. Equation 3 indeed indicates that the wind velocity is a key parameter governing the efficiency of the accretion.

Coming back to Fig. 5, the previous discussion can account for the lack of barium stars among giants with $[Fe/H] > -0.2$. In that framework, it would be necessary (and not unrealistic, except maybe for ξ^1 Ceti) to assume that all systems with normal giants of low-metallicity ($[Fe/H] < -0.2$) present in Fig. 5 actually have main sequence companions. They are not barium stars simply because they never experienced wind accretion. The system ξ^1 Ceti deserves a few more comments, however. In their discovery paper, Böhm-Vitense & Johnson (1985) note that rather unusual physical parameters were adopted for the WD companion of ξ^1 Ceti in an attempt to minimize the discrepancy between the observed and predicted UV fluxes. They suggested that absorbing dust or gas may be present in the system. But on the other hand, McAlister et al. (1984) reported speckle observations resolving the ξ^1 Ceti system, with an angular separation matching that predicted by the spectroscopic orbit (Griffin & Herbig 1981). This optical detection of the companion is puzzling, since a WD of $T_{eff} \sim 13\,000$ K as deduced by Böhm-Vitense & Johnson (1985) would be at least 10 mag fainter than the giant primary. More observations are clearly needed in order to clarify the nature of the companion.

5.   Conclusions

Binarity appears to be a necessary condition for producing barium stars. Wind accretion within the binary system is very likely to be responsible for the contamination of the barium star, as suggested by average eccentricities lower, on the average, than those of normal giants and by the correlation between the orbital period and the level of chemical peculiarities. Both effects can be used to constrain future numerical models of wind accretion in binary systems.

The existence of normal giants in binary systems with barium-like characteristics indicate that binarity is not sufficient to yield a barium star. Since barium stars are mainly found in moderately metal-poor populations, it is suggested that metallicity also plays a key role in the formation of the barium star. A low metallicity may either favour heavy element synthesis or dredge-ups in the former AGB companion, or increase the efficiency of wind accretion by lowering the AGB wind velocities.

Acknowledgments. We thank M. Mayor for communicating us unpublished CORAVEL data relative to several dwarf barium stars, as well as T. Theuns for useful comments on the manuscript.

6.   References

Abt, H.A., Levy, S.G., 1976, ApJS 30, 273

Barbuy, B., Jorissen, A., Rossi, S.C.F., Arnould, M., 1992, A&A, in press

Bartkevicius, A., Sviderskiene, Z., 1981, Bull. Vilnius Astron. Observ. 58, 54

Batten, A.H., Fletcher, J.M., MacCarthy, D.G., 1989, Publ. Dominion Astrophys. Observ. 17, 1

Bergeron, P., Saffer, R.A., Liebert, J., 1991, in: Confrontation between Stellar Pulsation and Evolution, eds. C. Cacciari & G. Clementini, PASP Conf. Ser. Vol. 11, p. 513

Bidelman, W.P., Keenan, P.C., 1951, ApJ 114, 473

Bidelman, W.P., 1981, AJ 86, 553

Bidelman, W.P., 1983, AJ 88, 1182

Bidelman, W.P., 1985, AJ 90, 341

Boffin, H.M.J., Jorissen, A., 1988, A&A 205, 155

Boffin, H.M.J., Cerf, N., Paulus, G., 1992, A&A, submitted

Böhm-Vitense, E., 1980, ApJ 239, L79

Böhm-Vitense, E., Nemec, J., Proffitt, Ch., 1984, ApJ 278, 726

Böhm-Vitense, E., Johnson, H.R., 1985, ApJ 293, 288

Bond, H.E., 1984, in: Future of Ultraviolet Astronomy based on Six years of IUE Research, eds. J.M. Mead, R.D. Chapman & Y. Kondo, NASA Conf. Publ. 2349, p.289

Bondi, H., Hoyle, F., 1944, MNRAS 104, 273

Brown, J.A., Smith, V.V., Lambert, D.L., Dutchover, E.Jr., Hinkle, K.H., Johnson, H.R., 1990, AJ 99, 1930

Burki, G., Mayor, M., 1983, A&A 124, 256

Catchpole, R.M., Robertson, B.S.C., Warren, P.R., 1977, MNRAS 181, 391

Cowley, A.P., 1976, PASP 88, 95

Cowley, C.R., Downs, P.L., 1980, ApJ 236, 648

Crawford, D.L., Mandwewala, N., 1976, PASP 88, 917

Dahn, C.C., Liebert, J., Kron, R.G., Spinrad, H., Hintzen, P.M., 1977, ApJ 216, 757

Dearborn, D.S.P., Liebert, J., Aaronson, M., Dahn, C.C., Harrington, R., Mould, J., Greenstein, J.L., 1986, ApJ 300, 314

Dominy, J.F., Lambert, D.L., 1983, ApJ 270, 180

Duquennoy, A., Mayor, M., 1991, A&A 248, 485

Eggen, O.J., 1972, MNRAS 159, 403

Eggleton, P.P., 1986, in: The Evolution of Galactic X-ray Binaries, eds. J. Truemper, W.H.G. Lewin & W. Brinkmann, Dordrecht: Reidel, p. 87

Fekel, F.C.Jr., Simon, T., 1985, AJ 90, 812

Forestini, M., Goriely, S., Jorissen, A., Arnould, M., 1992, A&A, in press

Gingold, R.A., Monaghan, J.J., 1977, MNRAS 181, 375

Green, P.J., Margon, B., MacConnell, D.J., 1991, preprint

Griffin, R.F., 1991, The Observatory 111, 29

Griffin, R.& R., 1981, MNRAS 193, 957

Griffin, R.F., Herbig, G.H., 1981, MNRAS 196, 33

Hakkila, J., 1989a, AJ 98, 699

Hakkila, J., 1989b, A&A 213, 204

Hakkila, J., 1990, AJ 100, 2021

Hakkila, J., McNamara, B.J., 1987, A&A 186, 255

Johnson, H.R., 1992, in: Evolutionary processes in interacting binary stars (IAU Symp. 151), eds. Y. Kondo, R. Polidan & R. Sistero, Dordrecht, Kluwer, in press

Johnson, H.R., Ake, T.B., Ameen, M.M., 1992, preprint

Jorissen, A., Mayor, M., 1988, A&A 198, 187

Jorissen, A., Arnould, M., 1989, A&A 221, 161

Jorissen, A., Manfroid, J., Sterken, C., 1991, The Messsenger 66, 53

Jorissen, A., Manfroid, J., Sterken, C., 1992a, A&A 253, 407

Jorissen, A., Smith, V.V., Lambert, D.L., 1992b, A&A, in press

Jorissen, A., Mayor, M., 1992, A&A, in press

Jura, M., 1988, ApJS 66, 33

Jura, M., 1991, A&A Rev. 2, 227

Käppeler, F., Beer, H., Wisshak, K., 1989, Rep. Prog. Phys. 52, 945

Kippenhahn, R., Meyer-Hofmeister, E., 1977, A&A 54, 539

Knapp, R., 1985, ApJ 293, 273

Knapp, R., Morris, M., 1985, ApJ 292, 640

Kollatschny, W., 1980, A&A 86, 308

Kovács, N., 1985, A&A 150, 232

Lambert, D.L., 1985, in: Cool Stars with Excesses of Heavy Elements, eds. M.
    Jaschek & P.C. Keenan, Dordrecht, Reidel, p.191

Lambert, D.L., 1988, in: The Impact of High S/N Spectroscopy on Stellar
Physics (IAU Symp. 132), eds. G. Cayrel de Strobel & M. Spite, Dordrecht: Kluwer,
p. 563

Lambert, D.L., Smith, V.V., Coleman, H., 1991, private communication

Lecar, M., Wheeler, J.C., McKee, C.F., 1976, ApJ 205, 556

Little-Marenin, I.R., Little, S.J., 1987, AJ 93, 1539

Little, S.J., Little-Marenin, I.R., Hagen-Bauer, W., 1987, AJ 94, 981

Livio, M., Warner, B., 1984, The Observatory 104, 152

Lü, P.K., 1991, AJ 101, 2229

Lü, P.K., Upgren, A.R., 1979, AJ 84, 101

Lü, P.K., Dawson, D.W., Upgren, A.R., Weis, E.W., 1983, ApJS 52, 169

Luck, R.E., Bond, H.E., 1991, ApJS 77, 515

Lucy, L., 1977, AJ 82, 1013

MacConnell, D.J., Frye, R.L., Upgren, A.R., 1972, AJ 77, 384

McAlister, H.A., Hartkopf, W.I., Gaston, B.J., Hendry, E.M., Fekel, F.C., 1984, ApJS
    54, 251

McClure, R.D., 1983, ApJ 268, 264

McClure, R.D., 1984a, PASP 96, 117

McClure, R.D., 1984b, ApJ 280, L31

McClure, R.D., 1985, in: Cool Stars with Excesses of Heavy Elements, eds. M.
    Jaschek & P.C. Keenan, Dordrecht, Reidel, p. 315

McClure, R.D., 1989, in: Evolution of peculiar red giant stars (IAU Coll. 106), eds.
    H.R. Johnson & B. Zuckerman, Cambridge U.P., p. 196

McClure, R.D., van den Bergh, S., 1968, AJ 73, 313

McClure, R.D., Fletcher, J.M., Nemec, J.M., 1980, ApJ 238, L35

McClure, R.D., Woodsworth, A.W., 1990, ApJ 352, 709

McWilliam, A., 1988, Ph.D. thesis, Univ. of Texas at Austin

McWilliam, A., 1990, ApJS 74, 1075

Meyer, F., Meyer-Hofmeister, H., 1979, A&A 78, 167

Morbey, C.L., Griffin, R.F., ApJ 317, 343

Nandy, K., Thompson, G.I., Jamar, C., Monfils, A., Wilson, R., 1976, A&A 51, 63

Neo, S., Miyaji, S., Nomoto, K., Sugimoto, D., 1977, PASJ 29, 249

North, P., Duquennoy, A., 1991, A&A 244, 335

North, P., Lanz, T., 1991, A&A 251, 489

Paczyński, B., 1971, ARA&A 9, 183

Paczyński, B., 1976, in: Structure and Evolution of Close Binary Systems (IAU Symp. 73), eds. P.P. Eggleton, S. Mitton & J.A.J. Whelan, Dordrecht, Reidel, p. 75

Popper, D.M., 1989, ApJS 71, 595

Proffitt, C.R., 1989, ApJ 338, 990

Proffitt, C.R., Michaud, G., 1989, ApJ 345, 998

Reimers, D., 1984, A&A 136, L5

Richer, H.B., 1989, in: Evolution of peculiar red giants (IAU Coll. 106), eds. H.R. Johnson & B. Zuckerman, Cambridge U.P., p. 35

Scalo, J.M., 1976, ApJ 206, 474

Schindler, M., Stencel, R.E., Linsky, J.L., Basri, G.S., Helfand, D.J., 1982, ApJ 263, 269

Simon, T., Fekel, F.C.Jr., Gibson, D.M., 1985, ApJ 295, 153

Smith, V.V. 1984, A&A, 132, 326

Smith, V.V., Lambert, D.L., 1986, ApJ 303, 226

Smith, V.V., Lambert, D.L., 1987a, MNRAS 226, 563

Smith, V.V., Lambert, D.L., 1987b, AJ 94, 977

Smith, V.V., Lambert, D.L., 1988, ApJ 333, 219

Smith, V.V., Lambert, D.L., 1990, ApJS 72, 387

Strassmeier, K.G., Hall, D.S., Zeilik, M., Nelson, E., Eker, Z., Fekel, F.C., 1988, A&AS 72, 291

Tassoul, J.L., 1988, ApJ 324, L71

Tomkin, J., Lambert, D.L., 1983, ApJ 273, 722

Tomkin, J., Lambert, D.L., 1986, ApJ 311, 819

Tomkin, J., Lambert, D.L., Edvardsson, B., Gustafsson, B., Nissen, P.E., 1989, A&A 219, L15

Tout, C.A., Eggleton, P.P., 1988, MNRAS 231, 823

Tout, C.A., Hall, D.S., 1991, MNRAS 253, 9

Utsumi, K., 1985, in: Cool Stars with Excesses of Heavy Elements, eds. M. Jaschek & P.C. Keenan, Dordrecht, Reidel, p.243

Wannier, P.G., Sahai, R., Andersson, B.G., Johnson, H.R., 1990, ApJ 358, 251

Warner, B., 1965, MNRAS 129, 263

Webbink, R.F., 1986, in: Critical Observations vs Physical Models for Close Binary Systems, eds. K.C. Leung and D.S. Zhai, New York, Gordon and Breach

Zahn, J.P., 1966, Ann. d'Astrophys. 29, 313

Zahn, J.P., 1975, A&A 41, 329

Zahn, J.P., 1977, A&A 57, 383

Zahn, J.P., 1989, A&A 220, 112

Zuckerman, B., Dyck, H.M., 1989, A&A 209, 119

Spectroscopic Binaries in the Open Cluster M67

David W. Latham *   Robert D. Mathieu †
Alejandra A. E. Milone ‡   Robert J. Davis §

Abstract

For almost 400 members of the old open cluster M67 we have accumulated about 5,000 precise radial velocities. Already we have orbital solutions for 34 spectroscopic binaries. Many of these orbits were derived by combining Palomar and CfA observations, thus extending the time coverage to more than 20 years. The distribution of eccentricity versus period shows evidence for tidal circularization on the main sequence. The transition from circular orbits is fairly clean. Excluding the blue stragglers, the first eccentric orbit has a period of 11.0 days, while the last circular orbit has a period of 12.4 days. For longer periods the distribution of eccentricity is the same as for field stars. The blue straggler S1284 has an eccentric orbit despite its short period of 4.2 days.

1.   Radial Velocities of M67 Members

In 1971 Roger Griffin and Jim Gunn began monitoring the radial velocities of most of the members brighter than the main-sequence turnoff in the old open cluster M67, primarily using the 200-inch Hale Telescope. In 1982, just as the sequence of observations at Palomar was ending, Dave Latham and Bob Mathieu began monitoring many of the same stars with the 1.5-m Tillinghast Reflector and the Multiple-Mirror Telescope on Mount Hopkins. The Palomar and Mount Hopkins data sets were successfully merged, together with some additional CORAVEL velocities kindly provided

*Harvard-Smithsonian Center for Astrophysics, 60 Garden Street, Cambridge, Massachusetts 02138, USA

†Department of Astronomy, University of Wisconsin, 475 N. Charter Street, Madison, Wisconsin 53706, USA

‡Harvard-Smithsonian Center for Astrophysics, 60 Garden Street, Cambridge, Massachusetts 02138, USA, and Córdoba Observatory, National University of Córdoba, Laprida 854, 5000 Córdoba, Argentina

§Harvard-Smithsonian Center for Astrophysics, 60 Garden Street, Cambridge, Massachusetts 02138, USA

by Michel Mayor, to obtain 20 years of time coverage (e.g. see [MLG86]). Based on these data, orbital solutions have been published by Mathieu et al.[MLG90] for 22 spectroscopic binaries, down to a magnitude limit of $V = 12.7$. This magnitude limit reaches into the main-sequence turnoff in M67, but does not extend safely onto the main sequence itself. The stars at the main-sequence turnoff in M67 have masses of about 1.2 solar masses, which is close to the dividing line where the more massive stars do not have significant convective zones. In order to reach well onto the main sequence, to masses where the stars have convective zones, the M67 survey is now being extended to $V = 15.5$ with the telescopes at Mount Hopkins. Already we have 13 additional orbital solutions, with the promise of many more to come.

2.   Binaries in M67

In Table I we list the period and eccentricity for 34 spectroscopic binaries which are members of M67. For the binaries with circular orbits, Mathieu et al.[MLG90] adopted an eccentricity of 0.0. Here we have allowed the eccentricity to be a free parameter in the orbital solution, so that the estimated error in the eccentricity may be used to judge whether the eccentricity is significantly different from circular. The new orbital solutions are denoted by [CfA] in the reference column of Table I.

The stars in Table I are listed in order of decreasing brightness, using the V magnitudes and $B - V$ colors from Girard et al. [GGL89]. This allows the systems with evolved primaries to be distinguished from the systems with main-sequence primaries. The designations for the stars listed in Table I are taken from Sanders [San77].

3.   Eccentricity versus Period Distribution

In Figure 1 we plot eccentricity versus log period for the M67 binaries that have orbital solutions. All the binaries with periods shorter than 11 days have circular orbits, presumably due to tidal circularization (cf. [Zah77], [MaM88], [MDL92]). The transition from circular to eccentric orbits is fairly clean. The last circular orbit has a period of 12.45 days, somewhat longer than the period of 11.02 days for the first eccentric orbit.

The ten M67 binaries with periods longer than 12.4 days show a wide spread of eccentricities, with a mean eccentricity of 0.35 ± 0.07. This is somewhat smaller than the mean eccentricity of 0.42 ± 0.03 for a sample of 44 main-sequence binaries with periods longer than 14 days found among 164 nearby solar-type field dwarfs [DuM91]. Two of the main-sequence binaries in M67 with fairly long periods have nearly circular orbits: S1292 with period 100 days and S1247 with period 70 days. Perhaps these two binaries were formed with nearly circular orbits, or perhaps tidal circularization has been more effective due to an extensive convective zone in the companion, if it has already gone through its post-main-sequence evolution and is

Table 1. M67 binaries

Name	V	B − V	Period	Eccentricity			Reference
S1250	9.71	1.35	4410.000	0.50	±	0.03	[MLG90]
S1221	10.76	1.13	6445.000	0.031	±	0.027	[CfA]
S1237	10.78	0.94	697.800	0.105	±	0.015	[MLG90]
S1284	11.04	0.22	4.1828	0.266	±	0.045	[Mil91]
S1072	11.32	0.63	1495.000	0.32	±	0.07	[MLG90]
S1040	11.52	0.88	42.8271	0.027	±	0.028	[CfA]
S1264	11.74	0.92	353.900	0.383	±	0.013	[MLG90]
S1053	12.19	0.84	123.39	0.486	±	0.018	[MLG90]
S2206	12.30	0.84	18.377	0.341	±	0.065	[MLG90]
S1272	12.52	0.65	11.0215	0.262	±	0.007	[MLG90]
S1045	12.55	0.59	7.64521	0.007	±	0.005	[CfA]
S251	12.57	0.67	948.000	0.554	±	0.024	[MLG90]
S1285	12.59	0.65	277.800	0.19	±	0.03	[MLG90]
S999	12.60	0.78	10.05525	0.014	±	0.020	[CfA]
S1234	12.66	0.59	4.35563	0.058	±	0.023	[MLG90]
S1000	12.69	0.88	531.1	0.092	±	0.018	[MLG90]
S1242	12.70	0.72	31.7797	0.664	±	0.018	[MLG90]
S1024	12.72	0.57	7.15961	0.005	±	0.005	[CfA]
S1508	12.73	0.70	25.866	0.441	±	0.007	[MLG90]
S986	12.74	0.56	10.3386	0.006	±	0.008	[CfA]
S1216	12.75	0.56	60.445	0.451	±	0.007	[MLG90]
S821	12.85	0.55	584.16	0.793	±	0.012	[CfA]
S1292	13.24	0.62	100.41	0.034	±	0.028	[CfA]
S990	13.45	0.55	41.847	0.352	±	0.018	[CfA]
S973	13.49	0.56	40.420	0.199	±	0.020	[CfA]
S1063	13.52	1.07	18.3874	0.217	±	0.014	[CfA]
S1009	13.70	0.56	5.9531	0.014	±	0.015	[CfA]
S1224W	13.7		12.448	0.059	±	0.030	[CfA]
S1070	13.98	0.61	2.66061	0.010	±	0.015	[CfA]
S1247	14.05	0.61	69.77	0.056	±	0.010	[CfA]
S1014	14.11	0.81	16.219	0.307	±	0.011	[CfA]
S981	14.14	0.62	55.831	0.508	±	0.047	[CfA]
S2222	14.76	0.82	25.6712	0.544	±	0.011	[CfA]
S810	15.16	0.92	10.151947	0.014	±	0.021	[CfA]

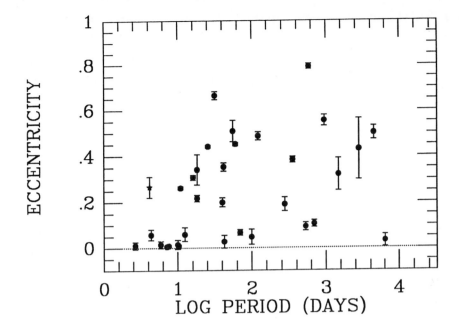

Figure 1. Eccentricity versus log period for 34 binaries in M67

now a white dwarf. It might be possible to confirm this suggestion by checking for evidence that the companions of S1292 and/or S1247 are white dwarfs.

4.   The Blue Straggler S1284

The old open cluster M67 is one of the first stellar systems where blue stragglers could be identified ([JoS55], [EgS64]). For 9 of the 13 "classical" blue stragglers in M67, we are able to measure radial velocities with the CfA digital speedometers. Although 5 of these blue stragglers show definite velocity variations in our data, S1284 (=F190 [Fag06]) is the only one with a short period [MiL92]. This makes S1284 especially interesting as a possible candidate for a history of mass transfer, one of the classical mechanisms proposed as a way to manufacture blue stragglers in an old stellar population [McC64].

There can be no doubt that S1284 is a member of M67. The 1950 position (08:48:50.18 +12:02:28.3 [GGL89]) places it 3 arc min from the center of the cluster, which has a half-mass radius for the single stars of about 10 arc min [MaL86]. In addition, the proper motion of the star matches the cluster motion very closely, with a 99 percent probability of membership on this basis [GGL89]. Finally, the   velocity of the spectroscopic orbit is 34.42 ± 0.48 km s^{-1}, which matches the cluster velocity of 33.6 km s^{-1} within the errors and cluster velocity dispersion of 0.5 km s^{-1} [MiL92].

Although S1284 had long been suspected of being a spectroscopic binary with a period of about 4 days [Deu68], it is only recently that an orbit has been solved and published ([MLK91], [MiL92]).

Has S1284 been turned into a blue straggler by recent mass transfer? Milone and Latham [MiL92] have explored a scenario in which S1284 was originally a close binary composed of two main-sequence stars [CoJ84]. The more massive component evolved first, expanded until it filled its Roche lobe, and then transferred mass onto its companion very quickly until the masses were equal. During this phase of mass transfer the orbital separation and period both became smaller. Subsequently, the mass transfer went much more slowly, because the orbital separation grew larger as the mass-donating star became progressively less massive than the mass-receiving star, and further evolution of the donating star was required to sustain the mass transfer. Milone and Latham [MiL92] conclude that S1284 has undergone efficient mass transfer, and that very little mass has been lost from the system.

However, the orbital eccentricity for S1284 is significant and remains a puzzle. When S1284 first started its mass transfer, its orbital period must have been very short, and therefore the orbit must have been circularized by tidal effects. For example, if no mass has been lost from the system, then the initial period would have been no longer than about a day [Pri85]. Such an orbit would be close to contact and would circularize very quickly. If the initial eccentricity is small, then mass transfer tends to circularize the orbit [Pac71], and we would thus expect S1284 to have a circular orbit now. Perhaps the eccentricity that we derive is an artifact caused by the same sort of line asymmetries, similar to the ones that can lead to fictitious eccentricities in the orbital solutions for Algols [Pop89]. Another possibility is that an accretion disk formed during the mass transfer. It has been shown recently that resonant phenomena in accretion disks can increase the orbital eccentricity of a binary [ACL91]. Yet a third possibility is that a distant third star in a wide orbit has pumped up some eccentricity. Simulations of such a configuration for S1284 suggest that an eccentricity of 0.2 can be produced, but only under rather special conditions [MaK91].

References

[ACL91] Artymowicz, P., Clarke, C. J., Lubow, S. H., Pringle, J. E. *The effect of an external disk on the orbital elements of a central binary.* 1991, ApJ, 370, L35

[CoJ84] Collier, A. C., Jenkins, C. R. *Close binary stars and old stellar populations: the blue straggler problem revisited.* 1984, MNRAS, 211, 391

[Deu68] Deutsch, A. J. *1968, Carnegie Institution of Washington Year Book 67 1967-68 (Baltimore: Garamond/Pridemark), p. 24*

[DuM91] Duquennoy, A., Mayor, M. *Multiplicity among solar-type stars in the solar neighborhood. II. Distribution of the orbital elements in an unbiased sample.* 1991, AA, 248, 485

[EgS64] Eggen, O. J., and Sandage, A. R. *New photoelectric observations of stars in the old galactic cluster M67.* 1964, ApJ, 140, 130

[Fag06] Fagerholm, E. Ueber den sternhaufen Messier 67. *1906, Inaugural Dissertation, Uppsala*

[GGL89] Girard, T. M., Grundy, W. M., Lopez, C. E., van Altena, W. F. *Relative proper motions and the stellar velocity dispersion of the open cluster M67.* 1989, AJ, 98, 227

[JoS55] Johnson, H. L., Sandage, A. R. *The galactic cluster M67 and its significance for stellar evolution.* 1955, ApJ, 121, 616

[MaL86] Mathieu, R. D., Latham, D. W. *The spatial distribution of spectroscopic binaries and blue stragglers in the open cluster M67.* 1986, AJ, 92, 1364

[MaK91] Mazeh, T., Krymolowski, Y. *1991 private communication.*

[MaM88] Mathieu, R. D., Mazeh, T. *The circularized binaries in open clusters: a new clock for age determination.* 1988, ApJ, 326, 256

[McC64] McCrea, W. H. *Extended main sequence of some stellar clusters.* 1964, MNRAS, 128, 147

[MDL92] Mathieu, R. D., Duquennoy, A., Latham, D. W., Mayor, M., Mazeh, T., Mermilliod, J.-C. *The distribution of cutoff periods with age: an observational constraint on tidal circularization theory.* 1992, this volume

[Mil91] Milone, A. A. E. *Los blue stragglers de M67 y otros cumulos abiertos.* 1991, PhD thesis, Córdoba

[MiL92] Milone, A. A. E., Latham, D. W. *The blue straggler F190: a case for mass transfer.* 1992, in IAU Symposium No. 151. Evolutionary Processes in Interacting Binary Stars, ed. Y. Kondo, R. F. Sister", and R. S. Polidan (Dordrecht, Kluwer Academic Publishers) in press

[MLG90] Mathieu, R. D., Latham, D. W., Griffin, R. F. *Orbits of 22 spectroscopic binaries in the open cluster M67.* 1990, AJ, 100, 1859

[MLG86] Mathieu. R. D., Latham, D. W., Griffin, R. F., Gunn, J. E. *Precise radial velocities of late-type stars in the open clusters M11 and M67.* 1986, AJ, 92, 1100

[MLK91] Milone, A. A. E., Latham, D. W., Kurucz, R. L., Morse, J. A. *Binaries among the blue stragglers in M67*. 1991, in The Formation of Star Clusters, ASP Conf. Ser., Vol. 13, ed. K. Janes (San Francisco: Astron. Soc. Pacific), p. 424

[Pac71] Paczyński, B. *Evolutionary processes in close binary systems.*   1971, ARA&A, 9, 183

[Pop89] Popper, D. E. *Radial velocities in 12 Algol binaries.* 1989, ApJS, 71, 595

[Pri85] Pringle, J. E. *Introduction.*   1985, in Interacting Binary Stars, eds.J. E. Pringle R. A. Wade (Cambridge, Cambridge Univ. Press), ch. 1

[San77] Sanders, W. L. *Membership of the old open cluster M67.* 1977, AAS, 27, 89

[Zah77] Zahn, J.-P. *Tidal friction in close binary stars.* 1977, AA, 57, 383

Spectroscopic Binaries in the Halo

David W. Latham [*]   *Tsevi Mazeh* [†]   *Guillermo Torres* [‡]

Bruce W. Carney [§]   *Robert P. Stefanik*[¶]   *Robert J. Davis* [‖]

Abstract

For almost 1500 stars in the Carney-Latham survey of proper-motion stars we have accumulated about 20,000 precise radial velocities. Already we have orbital solutions for more than 150 spectroscopic binaries in this sample, and about 100 additional binary candidates with variable velocity. We find that among the metal-poor halo field stars in this sample the frequency of short-period spectroscopic binaries is indistinguishable from that of the disk. The distribution of eccentricity versus period shows evidence for tidal circularization on the main sequence. For the binaries more metal poor than $[m/H] = -1.6$ there is a clean transition from circular to elliptical orbits at a period of about 19 days. For longer periods the distribution of eccentricity is the same as for stars in the disk of the Galaxy.

1.   Frequency of Halo Binaries

To study the orbital characteristics of the binaries in the old, metal-poor, slowly-rotating halo population of our galaxy, we turn to the nearby halo field stars, which just happen to be passing through the solar neighborhood. The Carney & Latham

[*]Harvard-Smithsonian Center for Astrophysics, 60 Garden Street, Cambridge, Massachusetts 02138, USA

[†]School of Physics and Astronomy, Raymond and Beverly Sackler Faculty of Exact Science, Tel Aviv University, Tel Aviv 69978, Israel, and Harvard-Smithsonian Center for Astrophysics, 60 Garden Street, Cambridge, Massachusetts 02138, USA

[‡]Harvard-Smithsonian Center for Astrophysics, 60 Garden Street, Cambridge, Massachusetts 02138, USA, and Córdoba Observatory, National University of Córdoba, Laprida 854, 5000 Córdoba, Argentina

[§]Department of Physics and Astronomy, University of North Carolina, Chapel Hill, North Carolina 27599-3255, USA

[¶]Harvard-Smithsonian Center for Astrophysics, 60 Garden Street, Cambridge, Massachusetts 02138, USA

[‖]Harvard-Smithsonian Center for Astrophysics, 60 Garden Street, Cambridge, Massachusetts 02138, USA

[CaL87] proper-motion sample has proven to be a rich source of halo binaries. The radial velocities of the original sample of almost 1000 stars have now been monitored for more than a decade, supplemented more recently by an additional sample of nearly 500 stars. Orbital solutions have already been published for 80 of these stars (Latham et al. [LMC88], Latham et al. [LMS92]), with many more than this number of orbits yet to come.

For his thesis, Torres [Tor91] analyzed the frequency of spectroscopic binaries identified in the Latham-Carney proper-motion sample. Using the definition that a halo star must be more metal poor than $[m/H] = -1.2$ and should lag the solar orbit by more than $V = -150$ km s^{-1}, he found 463 halo stars in the sample, of which 88 had been flagged as binary or multiple systems. This leads to a frequency of $19\pm2\%$ for binaries with periods shorter than about 5 years and secondary masses larger than 0.1 of the primary mass. The corresponding frequency that he calculated for the thin disk stars in the sample is $18\pm4\%$. His calculation for the volume-limited Duquennoy & Mayor [DuM91] sample of 165 nearby solar-type dwarfs gives a frequency of $23\pm3\%$. We conclude that the frequency of spectroscopic binaries is indistinguishable for the halo and the disk of the Galaxy, despite significant differences in the chemical composition and kinematics between these two populations.

2.   Eccentricity Distribution for Halo Binaries

Mathieu & Mazeh [MaM88] proposed that the orbital period at which there is a transition from circular to eccentric orbits can be used as a kind of clock to date the relative ages of coeval samples of low-mass main-sequence binaries. The basic idea is that all the short-period binaries have had their orbits circularized by tidal mechanisms (e.g. see Zahn [Zah77], Mathieu et al. [MDL92]) and the transition between circular and eccentric orbits moves to longer periods as a coeval sample of binaries grows older.

Unfortunately, there is not yet good agreement on the details of the theory of tidal circularization. Thus, it is premature to attempt to derive an age of the Galaxy from the transition period observed for the oldest stars. Instead, we need observations to test the competing versions of the theory; we need orbital solutions for rich populations of short-period binaries in several different coeval samples covering a wide range of ages.

The eccentricity versus log period for all the halo binaries more metal poor than $[m/H] = -1.6$ is summarized in Table 1 and plotted in Figure 1. This metallicity corresponds to the peak of the distribution for the halo field stars and the halo globular clusters (e.g., see Laird et al. [LRC88]). According to Morrison, Flynn, and Freeman [MFF90] this metallicity also marks the extreme end of the disk population. Thus, there should be little danger that the binaries plotted in Figure 1 are contaminated by disk stars. The orbit for one of these stars (HD 89499) was published by Ardeberg & Lindgren

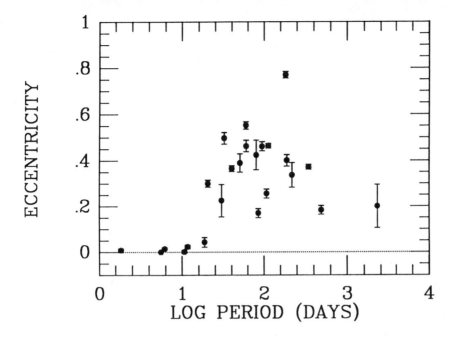

Figure 1. Eccentricity versus log period for halo binaries with [m/H] < −1.6.

[ArL91].

The last circular orbit in this sample occurs at 18.74 days (G65-22). Unfortunately, there are only two orbits with periods near 20 days, and the number of binaries in this region of the diagram needs to be increased. Towards this goal we have undertaken a velocity survey of nearly 500 low-metallicity stars reported by Ryan [Rya89]. The observing strategy is designed to identify binaries with periods shorter than about 30 days, and then to accumulate enough observations to allow orbital solutions. The availability of substantial amounts of telescope time with the CfA digital speedometers together with flexible scheduling make it feasible to solve for a few dozen new orbits in just a few years. In addition, it appears that there may soon be new orbits available for several metal-poor binaries observed with the southern CORAVEL (Ardeberg, Lindgren, & Lundström [ALL91]).

The halo binaries plotted in Figure 1 with periods longer than 19 days show a wide spread of eccentricities, with a mean of 0.38±0.04. This is similar to the mean eccentricity of 0.43±0.04 found for 36 Hyades binaries with periods longer than 10 days (Stefanik & Latham [StL92]), and of 0.42±0.03 for a sample of 44 binaries with period longer than 14 days found among 164 nearby solar-type field dwarfs (Duquennoy & Mayor [DuM91]). We conclude that the initial distribution of eccentricities for main sequence binaries must have been very much the same for the halo as for the disk, despite significant differences in the chemical composition and kinematics

Table 1. Binaries with [m/H] < −1.6

Name	Period	Eccentricity			[m/H]
BD +13°13	1.84422	0.007	±	0.008	−2.12
G183-9	6.20185	0.014	±	0.005	−1.75
HD 89499	5.57397	0			−1.78
G66-59	10.73673	0.001	±	0.007	−2.61
G176-27	11.7306	0.023	±	0.008	−2.12
G65-22	18.736	0.043	±	0.021	−1.74
G88-10	20.630	0.300	±	0.015	−2.66
G190-10	30.152	0.225	±	0.071	−1.96
G206-34	32.392	0.497	±	0.025	−2.89
BD +13°3683	40.263	0.365	±	0.013	−2.91
G87-45	50.67	0.390	±	0.040	−2.00
BD +28°2137	60.177	0.552	±	0.016	−2.16
BD +20°2030	60.593	0.463	±	0.025	−2.68
G178-27	80.81	0.424	±	0.064	−2.08
G103-50	85.91	0.170	±	0.020	−2.40
G62-40	95.451	0.462	±	0.020	−1.97
G86-40	108.48	0.256	±	0.020	−2.38
BD −3°2525	113.55	0.464	±	0.010	−2.29
BD +38°4955	183.85	0.771	±	0.014	−2.73
G89-14	190.31	0.401	±	0.025	−1.88
BD +17°4708	219.38	0.337	±	0.054	−1.95
BD +15°150	346.39	0.372	±	0.011	−1.77
G18-54	491.5	0.183	±	0.019	−1.86
G29-71	2270.	0.200	±	0.095	−2.40

between these two populations. The short-period halo binaries plotted in Figure 1 include three circular orbits with periods longer than 8.5 days, the upper limit that Zahn and Bouchet [ZaB89] quote for the transition period due to tidal circularization during the pre-main-sequence phase of stellar evolution. Perhaps these three binaries just happened to form with circular orbits. This seems unlikely, because none of the longer-period binaries show circular orbits. Or, perhaps these three binaries each formed with a more massive companion which has already gone through its post-main-sequence evolution, when tidal circularization would be much more effective due to the extended convective zones during the giant phase. The fact that two of these three binaries are double-lined systems argues that the companions have not yet undergone post-main-sequence evolution. Thus, it seems likely that the orbits of these three binaries were circularized on the main sequence, after they finished their pre-main-sequence evolution.

Acknowledgements. We thank Ed Horine, Jim Peters, Skip Schwartz, Dick McCrosky, Joe Caruso, and Joe Zajac for obtaining many of the 20,000 echelle spectra that have gone into this project.

References

[ArL91] Ardeberg, A., Lindgren, H. *Orbital elements for double stars of Population II. The system HD 89499.* 1991, AA, 244, 310

[ALL91] Ardeberg, A., Lindgren, H., Lundström, I. *The oldest stars.* 1991, ESO Messenger *63*, 37

[CaL87] Carney, B. W., Latham, D. W. *A survey of proper-motion stars. I. UBV photometry and radial velocities.* 1987, AJ, *93*, 116

[DuM91] Duquennoy, A., Mayor, M. *Multiplicity among solar-type stars in the solar neighborhood. II. Distribution of the orbital elements in an unbiased sample.* 1991, AA, 248, 485

[LRC88] Laird, J. B., Rupen, M. P., Carney, B. W., Latham, D. W. *A survey of proper-motion stars. I. The halo metallicity distribution function.* 1988, AJ, 96, 1908

[LMC88] Latham, D. W., Mazeh, T., Carney, B. W., McCrosky, R. E., Stefanik, R.P., Davis, R. J. *A survey of proper-motion stars. I. Orbits for 40 spectroscopic binaries.* 1988, AJ, 96, 567

[LMS92] Latham, D. W., Mazeh, T., Stefanik, R. P., Davis, R. J., Carney, B. W., Krymolowski, Y., Laird, J. B., Torres, G., Morse, J. A. *A survey of proper-motion stars. II. Orbits for the second 40 spectroscopic binaries.* 1992, AJ, in press

[MDL92] Mathieu, R. D., Duquennoy, A., Latham, D. W., Mayor, M., Mazeh, T., Mermilliod, J.-C. *The distribution of cutoff periods with age: an observational constraint on tidal circularization theory.* 1992, this volume.

[MLG90] Mathieu, R. D., Latham, D. W., Griffin, R. F. *Orbits of 22 spectroscopic binaries in the open cluster M67.* 1990, AJ, 100, 1859

[MaM88] Mathieu, R. D., Mazeh, T. *The circularized binaries in open clusters: a new clock for age determination.* 1988, ApJ, 326, 256

[MFF90] Morrison, H. L., Flynn, C., Freeman, K. C. *Where does the disk stop and the halo begin? Kinematics in a rotation field.* 1990, AJ, 100, 1191

[Rya89] Ryan, S. G. *Subdwarf studies. I. UBVRI photometry of NLTT stars.* 1989, AJ, 98, 1693

[StL92] Stefanik, R. P., Latham, D. W. *Binaries in the Hyades.* 1992, in Complementary Approaches to Double and Multiple Star Research, IAU Coll. 135, Eds H.A. McAlister and W.I. Hartkopf, in press

[Tor91] Torres, G. *Caracteristicas de las binarias espectroscopicas en el halo de la Galaxia.* 1991, PhD Thesis, Córdoba.

[Zah77] Zahn, J.-P. *Tidal friction in close binary stars.* 1977, AA, 57, 383

[ZaB89] Zahn, J.-P., Bouchet, L. *Tidal evolution of close binary stars. II. Orbital circularization of close binaries.* 1989, AA, 223, 112

Eccentricity Evolution of a Binary Embedded in a Disk

Stephen H. Lubow [*]   *Pawel Artymowicz* [†]

Abstract

We consider the gravitational interaction of an eccentric binary star system with a circumstellar and circumbinary disk. We analyze this interaction by performing disk simulations that include pressure, viscous, and tidal forces. Results can be partially understood by analytic considerations. Resonances provide the most powerful interaction and can cause rapid evolution of the binary eccentricity. For binaries with small eccentricity, $e < 0.2$, the resonance at $\Omega(r) = \Omega_b/3$ (with binary angular frequency Ω_b) dominates the disk response and causes the eccentricity to rapidly grow. For higher eccentricities $e \gtrsim 0.5$, there are many competing resonances which nearly cancel their effects on eccentricity growth. In this regime, the outcome is delicate and either weak growth or decay of binary eccentricity is possible.

For binary mass ratios less extreme than 5:1, there is a strong asymmetry between the resonance locations within a circumbinary versus a circumstellar disk. No resonant excitation occurs for the circumstellar disks with such mass ratios. Consequently, a circumstellar disk provides a much weaker contribution to eccentricity growth than does a circumbinary disk.

A pre-main sequence binary is likely to be surrounded by a disk. Our model suggests that binaries with small initial eccentricity acquire eccentricity $e \sim 0.5$ during their pre-main sequence evolution.

1. Introduction

There is now good observational evidence to support the existence of disks during binary pre-main sequence (PMS) evolution (see [Mat91] for review). It is generally accepted that single star formation must proceed through the formation of an accretion disk (see review by [SAL87]). It is also natural to believe that disks must be present in the PMS binary environment. For several models of binary star formation, disks are to be expected. In the continued fragmentation model [Bod81], [Bos86],

---

[*]Space Telescope Science Institute, Baltimore, MD USA
[†]Hubble Fellow, University of California, Santa Cruz, CA USA

binaries result as a collapsing cloud forms a disk or ring that hierarchically splits into several bodies. In the so-called disk-fission model [ARS91], a disk that surrounds a single PMS star undergoes a large-scale instability that causes a second star to form within. Another model is based on capture [CP90]. In this model, disks around single stars interact dissipatively to form a binary, with a remnant disk.

The gravitational interaction of a binary with a disk can effect the eccentricity evolution of the binary. Observational evidence is now accumulating to suggest that the binary eccentricity distribution (for all but possibly close binaries) is established during the PMS phase (see figure in [Mat92]). For binaries with periods longer than about 100 days, this distribution is fairly broad with eccentricities typically ranging from about 0.2 to 0.8 [DM90]. The issue we address here is whether this evidence can be understood through the effects of a binary-disk interaction during the PMS phase.

Naively, one might think that binary eccentricity damping must occur because the binary will transfer energy to the disk that is eventually dissipated. Since eccentricity is due to an excess of energy above that required for a circular orbit (for fixed angular momentum), binary circularization is to be expected. This simple argument of course neglects the angular momentum transfer between the binary and the disk.

Consider an eccentric binary that interacts with its circumbinary disk mainly at apastron (in the impulse approximation). The disk responds to the (rotating) binary tidal field by producing a distortion and a tidal lag due to the disk viscosity. The disk tidal lag produces a torque that reduces the binary angular momentum, resulting in eccentricity growth. (Binary energy changes due to the circumbinary disk also need to be considered, but can be shown to be less effective than angular momentum changes.) However, the most important interactions involve resonances within the disk. Resonant interactions require a very different analysis.

A rigorous foundation for the analysis of disk-satellite resonant interactions was established by Goldreich and Tremaine [GT81]. They analyzed the orbital evolution of a very small mass satellite that orbits a primary object surrounded by a disk. The satellite excites many resonances in the disk and carves out a very small gap about its orbit. Their conclusion was that the eccentricity is damped. However, two opposing mechanisms are in close competition, with damping being the net outcome by only a 5% margin. On the other hand, certain other effects (namely, resonant saturation) could change that conclusion.

Recently, this problem has been reinvestigated [ACLP91] in the context of binary star systems. The ingredients in this model are that the binary orbit is coplanar with the disk, and that the disk has viscosity and pressure. Disk self-gravity is ignored. The main new element here is that for binaries, the mass ratio is not very extreme, say 2 to 1. In the binary case, the mass distribution in the disk differs substantially from that assumed in the previous paragraph. A much larger gap is created in the disk (relative to the separation of the two objects). For binaries of relatively small eccentricity $e < 0.2$, there is a definite imbalance between the competing effects on eccentricity, with strong eccentricity growth as the outcome. Only one resonance plays a major role. Important issues remained from that study, some of which have been subsequently investigated and are reported here. We outline the main results; more details will be published elsewhere.

In section 2, we discuss the theory of disk resonances and our method of analysis. In section 3, we discuss disks for binaries in circular orbits. In section 4, we describe the case of small, nonzero binary eccentricity; in section 5 we discuss higher eccentricity. Section 6 discusses the circumstellar (inner) disk; section 7 contains a summary.

2.   Resonance Theory

We briefly outline some basic established facts concerning the resonant response of a disk to tidal forcing [GT81]. We adopt a polar coordinate system (r, θ) in the inertial frame that is centered on the binary center of mass. Consider a disk subject to a tidal potential of the form

$$\Phi(r, \theta, t) = \sum_{l,m} \phi_{l,m}(r) cos(m\theta - l\Omega_b t). \tag{1}$$

A binary with no eccentricity generates a tidal field that is static in the frame of the binary, where $\theta = \Omega_b t$. In that case, $\phi_{l,m} = 0$ for all $l \neq m$. In general, it can be shown that for $l \neq m$, $\phi_{l,m} \sim \phi_{m,m} e^{|l-m|}$, where e is the binary eccentricity. For small eccentricity, only resonances with small $|l-m|$ produce significant effects. Resonances will be termed nth order resonances if they are driven by potential component $\phi_{l,m}$ with $|l - m| = n$.

Away from resonances, in the absence of viscosity, disk fluid elements orbit the binary periodically in closed, tidally distorted orbits that result in no net exchange of energy or angular momentum with the binary. If the disk contains resonances, then a very powerful transfer of energy and angular momentum occurs. Two important classes of resonances need to be considered.

1. Lindblad resonances occur where $\Omega(r) = l\Omega_b/(m \pm 1)$. The zeroth order Lindblad resonances, located where $\Omega(r) = m\Omega_b/(m \pm 1)$, play a major role in defining the disk edges, at least for binaries of small eccentricity. First order Lindblad resonances which occur where $\Omega(r) = (m \mp 1)\Omega_b/(m \pm 1)$ cause eccentricity growth.

2. Corotational resonances occur where $\Omega(r) = l\Omega_b/m$. Under some circumstances, these resonances cause eccentricity to damp where $\Omega(r) = (m \pm 1)\Omega_b/m$ and will be called first order corotation resonances.

At such resonances, the fluid disk responds by generating waves that carry the angular momentum and energy deposited at the resonance by the binary. The binary responds through a secular change in its orbital elements. These lowest-order resonance locations are shown in Figure 1.

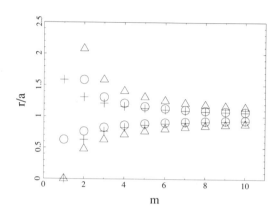

Figure 1. The resonance locations for a binary of extreme mass ratio in units of binary semimajor axis a versus angular harmonic m of the potential component that excites the resonance (see equation (1)). A circumbinary disk lies beyond $r/a = 1$; the circumstellar disk about the primary lies inside $r/a = 1$. Crosses are zeroth order Lindblad resonances, triangles are first-order Lindblad resonances, and circles are first order corotational resonances. (We omit resonances at $r = a$, which are of no interest for binaries.)

In spite of the fact that we can precisely locate these resonances and obtain formulae for the resonant torques on the disk, etc., a major ingredient must be fed into the theory that is not always easy to obtain. This factor is the disk mass density at each resonance. For fixed disk density, the effects of resonances become stronger for disk elements closer to the perturbing object. However, the disk density must be computed in a self-consistent fashion with the effects of the various resonances that shape the density distribution. Disk material closer to the perturber has a more powerful effect on the binary orbital evolution, but is subject to tidal truncation. The resonances close to the perturber that reside within the disk play an important role in determining the binary orbital evolution.

For very extreme mass ratios, such as for a sun-planetesimal system, one quite reasonably assumes that the disk mass distribution is smooth everywhere, with no gaps. For this case, complete analytic solutions can be determined for the orbital evolution of the small mass companion [War88], [Arty92]. For the binary case (or even the

Sun-Jupiter case), no such simple approximations can be applied. Instead, we adopt a numerical approach.

We use the method of smoothed particle hydrodynamics (SPH) to obtain the response of a disk to the effects of a binary of some prescribed eccentricity. The SPH scheme allows us to simulate the effects of gas pressure and viscosity. Once a quasisteady state is achieved in the disk, we determine the gravitational effects of the disk on the orbital elements of the binary. The Gauss equations [BC61] allow us to determine the time derivatives of the binary eccentricity from appropriately time-averaged gravitational disturbances created by disk particles.

We can partially check our results by applying the azimuthally-averaged mass distribution obtained from simulations to the resonance theory formulæ. The time derivative of eccentricity can then be determined from this mass distribution alone. This determination serves as an independent check of our application of the Gauss equations. We generally find agreement in these two methods at about the 30% level.

3. Zero Eccentricity

For a binary in a circular orbit, our numerical SPH results indicate that the circumbinary disk extends inwards to about $1.75a$, for binary semimajor axis (stellar separation) a. This location is in excellent agreement with the edge location expected from the tidal truncation model by Papaloizou and Pringle [PP77]. Paczynski and Rudak [PR80] have argued that the location of orbit crossings sets a lower limit on the inner radius of a circumbinary disk. For binaries, this radius occurs very close to the 1:2 outer resonance at $r \simeq 1.59a$, which is the radius for the outermost zeroth order (eccentricity independent) Lindblad resonance. Therefore, from several lines of argument we find a similar conclusion concerning the inner disk edge location about a circular binary. Notice that this disk edge location permits the outermost first order Lindblad resonance in Figure 1 to lie within the disk.

For a circumstellar disk, orbit crossings are such that the outermost extent of a disk is smaller than the 3:1 resonance location, where $\Omega = 3\Omega_b$ or $r = 0.48d$, for binary mass ratios less extreme than 5 to 1 [Pac77]. Since the innermost resonance in Figure 1 (apart from the zero-strength m=1 resonance that resides at the r=0) occurs at the 3:1 resonance, we do not expect to find strong resonant excitation of circumstellar disks for most observed binaries (see also [Lub92]).

4. Small Eccentricity

For small binary eccentricity, only first order resonances can be expected to dominate the eccentricity evolution. The nonzero eccentricity now causes the inner circumbinary disk truncation to retreat further away from the binary. In the case of small eccentricity $e = 0.1$ [ACLP91], the circumbinary disk is truncated slightly inside the

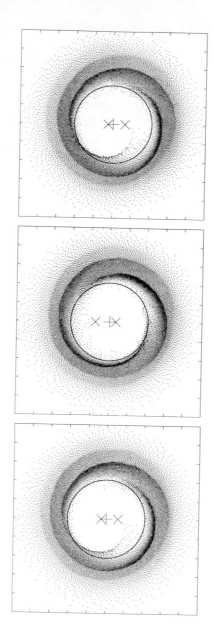

Figure 2. Plot of the circumbinary disk density response obtained from SPH simulations disk at 3 epochs. The cross denotes the binary center of mass; the Xs denote
the locations of the two stars having mass ratio 3:7. The circle denotes the location
of the 1:3 resonance. The middle frame is at one-half a binary orbit period later than
the top frame. The bottom frame is one orbit period later than the top frame. A
two-armed spiral, launched from the 1:3 resonance, rotates at half the binary angular
speed.

1:3 resonance, where $\Omega(r) = \Omega_b/3$. This resonance is the outermost resonance plotted in Figure 1 and is the only first order resonance that can lie within the disk. Since this resonance is of first order Lindblad type, eccentricity growth is expected, based on analytic theory. From our simulations, the resulting eccentricity growth time t_e for a binary with mass ratio 3:7 and eccentricity $e = 0.1$ is about

$$t_e = e/\,\dot{e} \simeq 9P/q_d, \tag{2}$$

where P is the binary orbital period. Disk mass fraction q_d is defined as $q_d = M_d/M_b$, where M_d the mass in circumbinary disk within $r = 6a$ of the binary center of mass with binary semimajor axis a, and M_b is the binary mass. For typically observed (nonclose) PMS binaries, P is of the order of years. If the mass M_d is related to the minimum solar nebula mass, then $q_d \sim 10^{-3}$ (the mass fraction of Jupiter). Therefore, $t_e \sim 10^4$ years is possible, considerably shorter than the estimated lifetimes of disks in T Tauri systems of several million years [Str87].

The theoretically expected disk response at this 1:3 Lindblad resonance is a launched two-armed, trailing, outwardly propagating spiral wave, having $l = 1$ and $m = 2$ in equation (1). The wave is predicted to be nonstationary in the frame of the binary and to rotate at half the angular speed of the binary. Figure 2 clearly shows these properties and the obvious dominance of this powerful resonant response.

5.   Higher Eccentricity

For binaries with higher eccentricity $e \gtrsim 0.5$, several effects come into consideration. First order resonances are more strongly excited, but second and higher order resonances come into play. As seen in Figure 3, the second order resonances having $|l - m| = 2$ can occur further out into the circumbinary disk, where densities may be higher. On the other hand, the higher eccentricity causes the circumbinary disk inner truncation also to occur further out. However, the most serious effect is that the excitation of these higher order resonances results in resonant contributions to eccentricity damping, as well as growth. Substantial cancellations of contributions occur.

For the resonances plotted in Figure 3, the innermost (outermost) set of Lindblad resonances in the circumbinary (circumstellar) disk cause eccentricity decay while the outermost (innermost) set of Lindblad resonances in the circumbinary (circumstellar) disk cause eccentricity growth.

For one particular model with $e = 0.5$, we find the inner edge of the circumbinary disk now occurs at $r \simeq 2.8a$. In general, for $e \gtrsim 0.5$, the eccentricity growth time is longer by about a factor of 100 relative to that of equation (2). Furthermore, for $e > 0.7$, eccentricity decay can actually occur depending on the exact details of the simulations. The cancellation of the various resonant effects becomes very

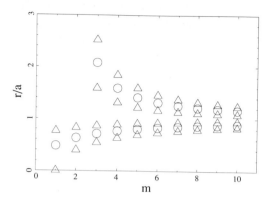

Figure 3. Same as Figure 1, but for second order resonances where $|l - m| = 2$.

delicate. Eccentricity growth during the PMS beyond eccentricities of 0.7 appears very unlikely, by the binary-disk interaction.

6.   Circumstellar Disk

As seen in Figure 1, there is a strong asymmetry in the resonance locations in a circumbinary versus circumstellar disk at small m. For small e, the zeroth order Lindblad resonances determine the edge locations. The outermost resonance in a circumbinary disk is evidently further away from its disk inner edge than is the innermost resonance in a circumstellar disk from its disk outer edge. This line of reasoning leads us to believe that eccentricity is much more easily excited in the circumbinary disk than the circumstellar disk.

Our SPH simulations bear out this expectation for a binary with $e = 0.1$ and mass ratio of 3:7. Only nonresonant interactions occur in circumstellar disks. Instead, some eccentricity growth occurs through a viscous binary-disk tidal interaction that was described in the introduction. However, the resulting growth rate is several orders of magnitude smaller than that due to the resonant interaction in the circumbinary disk.

7.   Summary

Our main conclusion is that resonant binary-disk interactions can substantially increase the eccentricities of typical (nonclose) binaries. We have argued that the interaction between a PMS binary and its surrounding disk is sufficiently strong to pump up the eccentricities to observed levels [Mat92] within the PMS phase. This mechanism greatly weakens at higher eccentricity ($e > 0.7$) and may set a natural limit for this process. The limit is not a sharp one, but does depend on details, such as the sound speed of the disk. The actual observed eccentricity distribution may then be expected to have some scatter, even if the initial binary eccentricities are uniformly small.

Although these results appear very promising in a statistical sense, the current model for binary GW Ori [MAL91] would appear to contradict our results, unless we happen to be observing it at a special time. GW Ori has a very small eccentricity $e \simeq 0.02$, and is the only PMS binary with period greater than 100 days and measured $e < 0.1$. This binary has a far-infrared eccess, indicative of a circumbinary disk. A dip in the spectral energy distribution has been interpreted as a gap in the disk that sets the inner edge of the circumbinary disk at about $3a$. There are several other unresolved issues about this system, such as the far-infrared energy source, that makes the current model of this system less than complete. The inner edge radius determination of $3a$ is somewhat model dependent, but for the purposes of discussion we regard it as a serious possibility. Our models indicate that a binary with the observed eccentricity of GW Ori would generate eccentricity about 10 times faster than the value of t_e of equation (2).

However, if we accept the suggested GW Ori model, where is the contradiction with our results that would predict rapid eccentricity growth? The contradiction is *not* with the resonant disk dynamics, but rather with the disk edge location. If the disk inner edge is really at $3a$, from Figure 1 we see that no resonant excitation of eccentricity is possible. However, none of our near-circular binary disk models have ever achieved a disk inner edge that lies at such a great distance from the binary. It is unclear what physical process sets an inner edge at that radius – no resonances can play a role there.

Much work remains to be done on this subject. Observed binary eccentricities, even among PMS stars, do not give us immediate information about the state of the binaries at formation, because of the strong binary-disk interaction during the PMS phase. With knowledge of this interaction, one may be able to place some constraints on the initial distribution of binary eccentricities. Only through such constraints can the observed eccentricities be used to reveal some information about the binary formation process.

ACKNOWLEDGEMENTS
We gratefully acknowledge continuing discussions on this problem with Cathie Clarke and James Pringle. I also thank Bob Mathieu for comments. P.A. acknowledges the support by Hubble Fellowship, NASA grant No. HF–1000.01–90A awarded by the Space Telescope Science Institute, which is operated by AURA for NASA.

References

[ARS91]   Adams, F. C., Ruden, S.P., and Shu, F.H. 1989, *Ap.J.* **347**, 959.

[Arty92] Artymowicz, P. 1992, preprint.

[ACLP91] Artymowicz, P., Clarke, C., Lubow, S.H., and Pringle, J.E. 1991, *Ap.J.* **370**, L35.

[Bod81]   Bodenheimer, P. 1981, in *IAU Symposium 93*, "Fundamental Problems in the Theory of Stellar Evolution", eds. D. Sugimoto, D. Q. Lamb and D. Schramm (Dordrecht: Reidel).

[Bos86]   Boss, A. P. 1986, *Ap. J. Suppl.* **62** , 519.

[BC61]   Brouwer, D., and Clemence G.M. 1961, *Celestial Mechanics*, (New York: Academic Press).

[CP90]   Clarke, C. J., and Pringle, J.E., 1990, *M.N.R.A.S*, **249**, 590.

[DM90]   Duquennoy, A., and Mayor M. 1990, in "Proc. XI European Astr. Meeting of the IAU", M.Vasquez ed., (Cambridge: University Press).

[GT81]   Goldreich, P., and Tremaine, S. 1981, *Ap. J.* **243**, 1062.

[Lub92]   Lubow, S.H. 1992, preprint.

[Mat91]   Mathieu, R. 1991, in *IAU Symposium 151*, Kondo, Sistero, and Polidan eds., preprint, (Dordrecht: Kluwer).

[Mat92]   Mathieu, R. 1992, this volume.

[MAL91]   Mathieu, R., Adams, F., and Latham, D.W. 1991, *A.J.*, **101**, 2184.

[Pac77]   Paczynski, B. 1977, *Ap.J.*, **216**, 822.

[PR80]   Paczynski, B. and Rudak, B. 1980, *Acta Astronomica* **31**, 13.

[PP77]   Papaloizou, J. C.B., and Pringle, J.E. 1977, *M.N.R.A.S* **181**, 441.

[SAL87]   Shu, F.H., Adams, F.C., and Lizano, S. 1987, *Ann. Rev. Astr. Ap.*, **25**, 23.

[Str87]   Strom, K.M., Strom, S., Edwards, S., Cabrit, S., and Skrutskie, M.F. 1989, *Astr. J.*, **97**, 1451.

[War88]   Ward, W. 1988, *Icarus* **73**, 330.

The Eccentricity Distribution of Pre-Main Sequence Binaries

*Robert D. Mathieu**

Abstract

The distribution of orbital eccentricity among pre-main sequence spectroscopic binaries is presented. Eccentricities from e=0 to e=0.8 are present by the pre-main sequence phase. The pre-main sequence and main-sequence orbital eccentricity distributions are very similar, except for a difference in the observed circularization cutoff periods. This similarity indicates that the main-sequence eccentricity distribution is largely established by ages of $\approx 10^6$ yr. Possible mechanisms establishing the eccentricity distribution are discussed, including pre-main sequence tidal circularization and disk-binary dynamics.

1. Introduction

The first pre-main sequence (PMS) binaries were discovered visually by Joy in the Taurus-Auriga clouds [JoB44], one of the nearest star-forming regions at a distance of 140 pc. At that distance, an angular separation of 0.25 arcsec (for the sake of argument) corresponds to 35 AU. A binary with two solar-mass stars and a semi-major axis of 35 AU has a period of 146 yr. Clearly either higher angular resolution or spectroscopic techniques are necessary to determine orbital elements for a large sample of PMS binaries. Observations of PMS stars with very high angular resolution have been obtained in the past decade with great success in the detection of binaries (e.g., [Zin90], [SCH92], [Ghe92]). As yet no orbits have been determined by these techniques, but they await only the passage of time. At present, our knowledge of the orbital eccentricity distribution of PMS binaries rests entirely upon spectroscopic orbital elements.

The first PMS spectroscopic orbit determination was that of V826 Tau [MWF83]. To date, orbital elements have been published for 11 PMS spectroscopic binaries. The author and collaborators have derived preliminary elements for an additional 8 PMS binaries in the Trapezium and NGC 2264 clusters and among naked T Tauri stars in the Orion association. The periods and eccentricities of these 19 PMS binaries are listed in Table 1. This list is not the product of a systematic survey of a well-defined sample of stars, although subsets do derive from such surveys [e.g., MWM89]. Nonetheless, the number of measured orbital eccentricities is now sufficient to permit a first analysis of the eccentricity distribution of binaries at ages of roughly 10^6 yr.

*Department of Astronomy, University of Wisconsin, Madison WI 53706 USA

Table 1: Pre-Main Sequence Spectroscopic Binaries

	P (d)	e	Ref.
155913-2233	2.4238	0	MWM
V4046 Sgr	2.43	0	BDLR
V826 Tau	3.88776	0	MR
ORI569	4.25	0	MWM2
EK Cep	4.42782	0.110	TPO
ORI429	7.46	0.30	MWM2
160905-1859	10.400	0.17	MWM
VSB126	12.924	0.18	MML
AK Sco	13.6093	0.469	AND
P1925	32.94	0.55	MML
P1540	33.73	0.12	MM
162814-2427	35.95	0.48	MWM
ORI477	74.1	0.42	MWM2
162819-2423S	89.1	0.41	MWM
ORI104A	93.4	0.66	MWM2
160814-1857	144.7	0.26	MWM
P1771	149.5	0.57	MML
GW Ori	242.	0.04 ± 0.06	MAL
VSB111	879.	0.80	MML

AND = [ALH89]; BDLR = [BY86], [DQT86]; MAL = [MAL91]; MM = [MaM88]; MML = [MML92]; MR = [MWF83], [RLN90]; MWM = [MWM89]; MWM2 = [MWM92]; TPO = [Tom83], [Pop87]

Clearly, one goal of such an analysis is to understand the origin of the main-sequence orbital eccentricity distribution. However, the study of the eccentricity distribution of young binaries may reflect more broadly on the star-formation process. The PMS eccentricity distribution is likely the result of numerous processes; a partial list would include binary formation, disk dynamics, stellar encounters and tidal circularization. Hence, the observed PMS eccentricity distribution may place constraints on essential star-formation issues as diverse as disk evolution in the presence of companions, the early evolution of stellar interiors, fragmentation of molecular cores, etc., as well as the binary formation process itself. In the course of this paper, a few examples of the potential of this line of study are given.[1]

---

[1] The observational study of PMS binaries is a rapidly opening forefront in the field of star formation. Reviews of other observational issues and results can be found in [Zin90], [BRM92] and [Mat92].

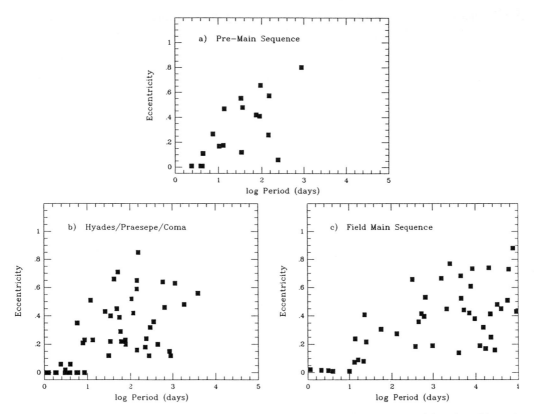

Figure 1. The period-eccentricity distribution for a) pre-main sequence binaries, b) main-sequence solar-mass binaries in the Hyades, Praesepe and Coma clusters [DMM92] and c) main-sequence solar-mass binaries in the field [DuM91].

2. The Eccentricity Distribution of Pre-Main Sequence Binaries

The distribution of orbital eccentricity with period for the PMS binaries listed in Table 1 is shown in Figure 1a. These data are compiled from many sources, and selection biases certainly exist. Nonetheless, several robust conclusions can be drawn.

First, orbital eccentricities from e=0 to e=0.8 have been established by the PMS phase of stellar evolution. The binaries listed in Table 1 typically have theoretical ages somewhat older than 10^6 yr. However, several of the binaries lie near the stellar birthline, indicating that they are among the youngest PMS stars. One such binary, P1925, is also one of the most eccentric binaries in the sample [MML92]. Evidently substantial orbital eccentricity is present in young binaries shortly after, and likely prior to, their unveiling at the stellar birthline.

Second, the morphology of the PMS eccentricity distribution has several notable features:

i) *All four binaries with periods less than 4.3 days have circular orbits.*

ii) *Circular or near-circular (e < 0.1) orbits are not found at longer periods.* The one notable exception is GW Orionis at a period of 242^d, a binary still embedded within a disk (Section 4).

iii) *The maximum orbital eccentricity increases with increasing period* to the longest period orbit at 879 days.

None of these features are likely due to selection effects. However, sampling statistics are a serious concern given the small numbers.

Third, there is no detectable difference between the PMS and main-sequence (MS) eccentricity distributions at periods where both have been observed; all of these features in the PMS eccentricity distribution are repeated in main-sequence eccentricity distributions. Figures 1b and 1c show two MS eccentricity distributions, taken from studies of the Hyades, Praesepe and Coma clusters [DMM92] and nearby field solar-mass stars [DuM91]. The qualitative similarity of the distributions is striking. Again the shortest period main-sequence orbits are typically circular, there are no near-circular orbits at longer periods and the maximum orbital eccentricity rises with increasing period to periods of 100^d to 1000^d. Formally, Kolmogorov-Smirnov tests show the two MS distributions to be indistinguishable from the PMS distribution. (Unfortunately there are only two field MS binaries with periods between 30^d and 300^d, presumably due to sampling statistics. Nonetheless, the maximum eccentricities in the PMS distribution in this period regime are similar to those expected from an interpolation between the MS eccentricity distributions at longer and shorter periods. For example, see Figure 3 in [Mat92] where the two distributions are overlaid.)

Quantitatively, one marked difference between the PMS and MS eccentricity distributions is the range in period of circular orbits. As noted, the longest period circular PMS binary has a period of 4.3^d. In contrast, the open cluster distribution has a circular orbit with a period of 8.5^d, and the field distribution shows only circular orbits up to a period of 10.3^d. The implied evolution with time of the longest period circular orbit is the focus of several papers in this volume, to which we leave further discussion (e.g., [MDL92], [DMM92]). However, in the next section we quantitatively compare the PMS short-period eccentricity distribution with theoretical prediction.

With this exception, the similarity of the distributions indicates that the MS orbital eccentricity distribution is largely established by $\approx 10^6$ yr. Thus, despite the youth of these binaries, the processes dictating their orbital eccentricities likely operated at a yet earlier stage in their evolution. Certainly substantial further orbital evolution in these PMS binaries is not required. Note, though, that at least half of the PMS binaries in this sample have no significant infrared excesses. As such, they do not have massive disks [SBW91] and any orbital evolution driven by such disks is completed [LuA92]. The eccentricity distribution of binaries still associated with massive disks may be quite different.

3. Early Evolution of Orbital Eccentricity: Pre-Main Sequence Tidal Circularization

The four shortest period PMS binaries all have circular orbits. The longest period circular orbit is 4.3^d, which can be taken to define the PMS circularization cutoff period [DMM92]. Circular orbits for the shortest period binaries are also found in main-sequence samples, but with longer cutoff periods. Typically these circular orbits have been attributed to tidal circularization. Whether that circularization occurs during the PMS phase, during the MS phase or during both is at present the focus of much discussion (e.g., many papers in this volume).

The rate of tidal circularization is very sensitive to the ratio of semi-major axis to stellar radius. At the stellar birthline, the radius of a solar-mass star is ≈ 5 times larger than at the main sequence. In addition, such a star is deeply convective. Both properties strongly promote rapid tidal circularization. Consequently, significant tidal circularization may occur during PMS evolution even though the phase is brief [MaM84]. Recently, Zahn and Bouchet [ZaB89; ZB] have convolved the tidal circularization theory of Zahn [Zah89] with the PMS stellar evolution models of Stahler, Shu and Taam to calculate the expected ZAMS circularization cutoff period for a variety of stellar mass combinations (0.5 M_O - 1.25 M_O). They find that tidal circularization during the PMS phase should produce circular orbits in ZAMS binaries with periods less than 7.3-8.5 days, and that no further circularization should occur on the main sequence.

This prediction can be directly tested with late-PMS or ZAMS binaries. In the ZB calculation most tidal circularization occurs while the stars are near the stellar birthline, prior to 10^5 yr for solar-mass stars. Thus, given the ZB models, the circularization process should be largely completed for the PMS binaries in Table 1. With one exception, the present sample of PMS binaries is consistent with the ZB prediction; unfortunately, though, only one PMS binary has been found with a period between approximately 5 days and 10 days. In this regard it is worth pointing out that in the Hyades and Praesepe clusters two binaries with periods of 5.8^d and 5.9^d have eccentricities of 0.35 [DMM92].

The one PMS binary contradicting the ZB prediction is EK Cep. This 4.4^d double-lined eclipsing binary contains a 2.0 M_O primary on the ZAMS with a 1.1 M_O secondary still contracting but very near to the ZAMS [Pop87]. The orbital eccentricity is 0.11, so taken straightforwardly EK Cep runs counter to the ZB prediction. However, this conclusion is subject to several caveats. First, the primary is more massive than has been considered by ZB, sufficiently so that it had a significantly different pre-main sequence evolutionary history [PaS90]. Second, the ZB calculations were made for binaries with initial eccentricities no higher than 0.3. Conceivably, EK Cep may have derived from a binary with much higher initial eccentricity. Third, the eccentricity of EK Cep is not large, and its source could be attributed to a tertiary companion [Maz91]. There is at present no evidence for such a tertiary. In summary, the binary EK Cep is not consistent with the ZB hypothesis, but as typical with a single case its significance as a test is not secure.

The theoretical prediction of a ZAMS cutoff at 7-8 days is also by no means secure among theorists. Others have argued for larger normalizations and differing period dependences for the rate of tidal circularization than those used by ZB (e.g., [GoM91], [DMM92]). Such alternative tidal circularization rates will likely produce different predictions for the efficacy of PMS tidal circularization and for the ZAMS cutoff period. Similarly, pre-main sequence stellar evolution theory is not firmly established, particularly for components of short-period binaries. The seminal calculations of ZB merit further exploration. As one example, a calculation has been made by Duquennoy et al. [DMM92] using a larger normalization and several alternative period dependences, but with main-sequence stellar structure. They find that a binary like EK Cep could derive from a MS binary with an initial period of 9^d after the passage of times comparable to the age of EK Cep.

4. Early Evolution of Orbital Eccentricity: Disks

There is strong observational evidence for extended material around many young binaries (see [Zin90] and [Mat92] for reviews). As with single stars the lack of large extinction toward these binaries suggests a disk-like distribution for this material, although the details of the spatial distribution (e.g., radial extent, thickness, etc.) are not at all well established for young binaries. Given sufficient mass a thin disk can produce substantial evolution in the orbital eccentricity of a binary embedded within the disk [LuA92]. Indeed, the structure and evolution of these disks may play a primary role in determining the main-sequence orbital eccentricities of many binaries.

The masses of disks around PMS stars can now be measured with millimeter and submillimeter observations; at these wavelengths disks are thought to be optically thin at most radii. Beckwith et al. [BSC89] surveyed a large sample of PMS stars at 1.3 mm and detected disk masses ranging from ≈ 0.001 M_O to ≈ 1 M_O, with an average near 0.03 M_O [BeS92]. These mass determinations may have a substantial systematic error due to the uncertainty in the mass opacity κ_v; for example, Adams, Emerson and Fuller [AEF90] use an opacity five times smaller than Beckwith et al. and consequently find larger masses. In either case these disk masses are sufficiently massive to cause orbital evolution in an embedded binary [LuA92].

The Beckwith et al. sample included 17 binaries. Among these binaries, they only found measurable disk masses for those binaries with projected separations greater than approximately 100 AU, a nominal radius for circumstellar disks around PMS stars. This intriguing result would suggest that the presence of an *embedded* binary may act dynamically to disperse disks, or that the formation process itself acts to consume or destroy disks. However, Beckwith and Sargent [BeS92] note that the result is not secure since their upper limits on the disk masses of shorter period binaries are comparable to many of their detections among the wider binaries. Also, binary detections at smaller separations are increasing in their sample. GG Tau, for which Beckwith et al. find a disk mass of 0.29 M_O, has since been found to be a binary with projected separation of 40 AU [LHR92]. Similarly, DQ Tau with a disk mass of 0.025 M_O is a candidate double-lined spectroscopic binary (Mathieu and Hartmann, in preparation). Finally, GW Orionis with a separation of 1 AU has been detected as a strong millimeter source (Weintraub, private communication). These anecdotal results suggest that disk masses sufficient to drive orbital evolution can be associated with binaries of separations less than 100 AU. In addition the typical detection limits of Beckwith et al. for binary disk masses (≈ 0.01 M_O) are large enough to miss dynamically interesting disks (Lubow, private communication). Thus it remains to be observationally established whether disks of sufficient mass to cause orbital evolution are typically present around PMS binaries with separations of less than 100 AU.

In this regard it is worth noting that circumbinary-disk-driven orbital evolution can be rapid [ACL91], so that for a binary with orbital period much less than its age the lack of a disk at present does not preclude earlier orbital evolution excited by a disk.

The star GW Orionis plays an important role in this discussion in that it is a PMS spectroscopic binary with many properties typically associated with the presence of a disk and with measured orbital elements. Mathieu, Adams and Latham [MAL91; MAL] discovered GW Ori to be a binary with a period of 242^d, a separation of approximately 1 AU and a near circular orbit. The star has an infrared excess from 1 µm to 1 mm, which MAL attributed to a thin disk. The millimeter flux suggests a substantial disk mass, although a measure of the mass has not yet been derived. The spectral energy distribution of GW Ori has structure in the infrared, with a minimum in the continuum around 10 µm and a very strong silicate emission feature. MAL were able to model the spectral energy distribution with a disk centered on the primary and having an annular gap between 0.2 AU and 3.3 AU, within which orbits the secondary star.

The near-circularity of the orbit at a period of 242 days is the most intriguing datum for the purposes of this paper.[2] Circular or near-circular orbits among main-sequence binaries with periods of greater than about ten days are found, but they are rare. Furthermore, recent numerical studies have found that a circumbinary disk (exterior to the binary orbit) rapidly drives eccentricity into near-circular binary orbits [ACL91]. A large increase in eccentricity has not occurred in the GW Ori system by an age $\approx 10^6$ yr, posing a challenge to the theory of binary-disk dynamics.

Lubow and Artymowicz [LuA92] note that the lack of eccentricity in the GW Ori orbit can be consistent with theory if there is a gap in the disk with an outer radius of 3 AU, as suggested by MAL. In this case, no disk mass resides at the strongest resonances and there is little coupling between the binary and the circumbinary disk. As they note, a physical mechanism for producing such a large gap is not known. An alternate but similar line of argument is to suggest that no circumbinary disk exists around GW Ori. As MAL point out, the far-infrared spectral energy distribution motivating the presence of a circumbinary disk can also be modeled by an extended shell-like distribution ($r_{min} > 100$ AU) of material reprocessing the stellar light. In this picture a circumbinary disk is permitted but not required by the observations. However, a circumstellar disk (interior to the binary orbit) and a large accretion rate remain required to explain the near-infrared excess, which would seem to make the presence of a circumbinary disk likely.

Several other PMS spectroscopic binaries also show infrared excesses suggesting disk material. Examples include 162814-2427, 162819-2423S, AK Sco, and V4046 Sgr. (Spectral energy distributions can be found in Figure 2 of [Mat92].) The periods of these systems are sufficiently short that the observed 10µm - 25µm excesses indicate material at distances greater than their orbital semi-major axes, e.g. in circumbinary disks. Submillimeter observations have been made for 162814-2427 and 162819-2423S [SBW91]; the upper limits do not rule out dynamically interesting disk masses.

With the exception of the very short-period binary V4046 Sgr, these binaries have

---

[2] As noted in MAL, however, the formal error on the eccentricity measurement is large and permits significantly higher values for the orbital eccentricity of GW Ori.

substantial orbital eccentricities. Evidently orbital eccentricity has been established prior to complete dispersal of their disks. A particularly intriguing case is the binary 162814-2427 with an orbital eccentricity of 0.48 at a period of 36^d. While in the near-infrared 162814-2427 has only an optically thin excess, the binary has an optically thick excess at 10 μm and 20 μm. Jensen *et al.* [JCM92] argue that no circumstellar disks are present and only a circumbinary disk remains (although a recently discovered tertiary at a projected separation ≈20 AU complicates this picture [Ghe92]). Perhaps this indicates that 162814-2427 is in a late stage of disk evolution, with the circumstellar disks having been exhausted through accretion [ACL91]. If GW Ori represents a typical early phase in binary-disk evolution and 162814-2427 a late phase, then these anecdotal cases hint that the early evolution of orbital eccentricity is intimately tied to the evolution of their associated disks.

5. Origins of Orbital Eccentricity: Discussion

Examination of Figure 1a clearly shows that the PMS eccentricity distribution depends on period. Thus different processes setting orbital eccentricity are likely at play in different period (and eccentricity) domains. Duquennoy and Mayor [DuM91] drew the same conclusion from consideration of the field MS eccentricity distribution. Here we consider the implications of the features of the PMS eccentricity distribution (Section 2) for the origins of orbital eccentricity.

5.1 Circular Orbits - P < 4.3^d

An essential result is that the shortest period binaries already have circular orbits by ages ≈10^6 yr. Arguably this confirms the conjecture of [MaM84] that tidal circularization will be rapid during the PMS phase of evolution, providing a straightforward explanation for these circular orbits.

We must be wary of limiting our ideas to those developed for detached main-sequence binaries, however; other processes are at play in the orbital evolution of young binaries. Early in the pre-main sequence phase the binaries may have circumbinary disks, whose influence has not yet been included in the circularization calculations. In addition, if accretion rates are comparable to single PMS stars the accreted mass over intervals as short as 10^5 yr can be dynamically significant; mass losses through outflows and winds are also present, with possible mass transfer effects. Furthermore the angular momentum evolution of the stars comes into play, possibly modified by the strong magnetic fields present. Finally, the orbital properties set by the formation process itself may not be lost by the PMS phase.

In the face of this array of processes, perhaps the most significant point is simply that circular orbits are only found for binaries with periods of a few days. This fact, in itself, may allow us to ascertain which processes are significant. The ZB calculation gives some preliminary evidence that such periods are those expected from PMS tidal circularization. As a second general point, if larger samples of PMS binaries continue to show only circular orbits at the shortest periods it would suggest that the circularity is established rather late in the orbital evolution, erasing the influence of both initial conditions and varying circumbinary environments.

Finally, *if* these binaries had similarly short periods at the stellar birthline, then their separations at that time were a few stellar radii or less. Given PMS binaries with periods as short as 2.4^d, it is difficult to avoid the conclusion that either the component stars were never at the stellar birthline or the orbital periods decreased during the PMS phase. The above processes change orbital period as well as eccentricity, so that the short periods as well as circular orbits may be important constraints. It is interesting that in some cases ZB found that tidal processes increase orbital period from the stellar birthline to the ZAMS.

5.2 Upper Envelope - $4.3^d < P < 879^d$

For periods above the circularization cutoff period, the PMS eccentricity distribution shows an upper envelope where the maximum eccentricity increases with increasing period. This trend is also evident in the MS eccentricity distributions up to periods of 100-1000 days. An alternative perspective is to note that for periods longer than the circularization cutoff period all three distributions show a triangular region in Figure 1 at short period and high eccentricity in which no binaries are found.

Periastron separation decreases with decreasing period and increasing eccentricity. Thus, for binaries with periods above the circularization cutoff, this triangular region would contain binaries with the smallest periastron separations. This is suggestive that interactions between binary components depopulate this region. An approximate scale length for such interactions can be obtained by plotting curves of constant periastron separation in the e-log P plane. In Figure 2 we show the PMS eccentricity distribution and a curve of constant periastron separation equal to $15 \ R_0$ for binaries having two solar-mass stars. The choice of $15 \ R_0$ was made to fit the shorter period envelope binaries. The curve is not a particularly good representation of the observed envelope, as it rises more steeply with increasing period. (Interestingly, this curve is an excellent fit to the upper envelope of the open cluster eccentricity distribution in Figure 1b.) Nonetheless the curve provides a fiducial mark for the periastron separation of binaries at the envelope. Recall that solar-mass stars at the stellar birthline have radii of approximately $5 \ R_0$. Thus the curve shows that binaries in much of the triangular "exclusion" region of the e-log P diagram would have periastron separations of less than a few stellar radii, *if* the component stars were at the stellar birthline. Indeed, over some of the region such stars would suffer physical collisions. At longer periods the observed envelope delineates wider periastron separations.

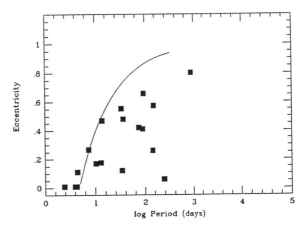

Figure 2. The pre-main sequence eccentricity distribution and the curve of constant periastron separation $r_{peri} = 15 \ R_0$.

During such close encounters at periastron passage dissipative interactions between the stars can be expected to cause rapid evolution of the orbits. The nature of the dissipation mechanisms (tides, disks, winds, magnetic fields) and their specific influences on the orbits need to be explored for explanation of the envelope. From a different perspective, analyses of the dissipation mechanisms may permit us to ascertain whether binaries along the upper envelope could remain there if they had stars at the stellar birthline. If not, again either orbital evolution (including capture) or different evolutionary tracks for binary components is indicated.

Certainly tidal effects would play a major role in the orbital evolution and likely are key to understanding the upper envelope. Indeed incomplete tidal circularization is a plausible explanation for the predominance of small eccentricities found at periods immediately above the circularization cutoff periods in both the PMS and MS samples. Furthermore, circularization should become progressively more incomplete with longer period, thus producing an envelope qualitatively similar to that observed. However, for a given binary the rate of tidal circularization is very sensitive to period. Hence the key issue is the longest periods at which significant tidal reduction of eccentricity can occur during the pre-main sequence phase. For PMS stars, tidal circularization is promoted by the stars' large radii and deep convection zones, but the available time is very short. We encourage that theoretical calculations of tidal circularization during the PMS phase of evolution (e.g., [ZaB89]) be extended to the study of incomplete circularization in PMS binaries of longer periods and higher eccentricities.

5.3 Absence of Near-Circular Orbits - $4.3^d < P < 879^d$

The lack of circular and near-circular orbits (e < 0.1) for periods longer than the circularization cutoff period is evident in both the PMS eccentricity distribution and the MS eccentricity distributions. This observation may be the signature of disks driving the orbital evolution. Theory argues that disks in a binary environment rapidly drive eccentricity into circular or low-eccentricity orbits. Lubow and Artymowicz [LuA92] suggest that eccentricity growth will be particularly strong at the smallest eccentricities (e < 0.2). If sufficiently massive disks are typically associated with young binaries, even if only early in their evolution, then the lack of circular and near-circular orbits among main sequence binaries would be a natural consequence. Indeed the presence of numerous binaries with orbital eccentricities as low as ≈0.2 may be an interesting constraint on this theory.

5.4 Eccentricity Distribution - $P \gtrsim 1000^d$

For periods longer than ≈1000^d discussion of the origin of orbital eccentricity must rely on the field main-sequence eccentricity distribution alone (Figure 1c). The lack of circular and near-circular orbits (e < 0.1) for periods longer than the circularization cutoff seen in the PMS eccentricity distribution remains evident in this period regime. At large eccentricities the observed distribution shows a marked lack of binaries for e > 0.8. As selection effects have not yet been evaluated, this observed decrease in frequency must be treated with caution.

As at shorter periods, the lack of near-circular orbits again is arguably attributable to disk excitation of eccentricity. Interestingly, Lubow and Artymowicz [LuA92] also argue that eccentricity excitation from PMS disks is unlikely above e≈0.7. If the binary formation process does not produce very high eccentricities, disks would not be able to drive binaries to such eccentricities. Thus, given a variety of disk structures and evolutionary histories a distribution of eccentricities between e≈0.2 and e≈0.7 would be expected, if not specified. Hence the true frequency of very high eccentricity orbits could be a key constraint on the role of disks.

At the very longest periods the influence of disks likely diminishes. A relevant scale length is the typical protostellar disk radius, perhaps of order 100 AU (or $P \approx 10^5$ days). The details of the orbital dynamics (and likely the formation process) must differ between binaries with separations much greater than or less than this scale length. (See [LuA92] for comparison of the dynamics of circumstellar and circumbinary disks.) Numerous other mechanisms for setting orbital eccentricity are certainly available. Fragmentation calculations produce eccentric orbits very early on, although the consequent evolution of those orbits has not been calculated. Duquennoy and Mayor [DuM91] argue that the eccentricity distribution of their longest period binaries tends toward the dynamically relaxed distribution f(e)=2e, which they suggest may be indicative of dynamical processes dictating the orbital eccentricity distribution (e.g., encounters in small groups of stars). In a similar vein Clarke and Pringle [ClP91] have argued that encounters of several single stars having disks can also produce binaries with a wide range of periods and orbital eccentricities.

These processes can also be important for periods less than 10^5 days; their relative importance, along with disks, remains to be assessed at all periods. Note though that in some cases different processes may be exclusive (e.g., capture formation of binaries destroys circumbinary disks). Attempting to isolate specific processes is a necessary simplification at this early stage, but in fact the eccentricity distribution, including the circularization cutoff period, is likely set by an interplay between tidal circularization, disk dynamics, encounters and other processes perhaps including binary formation itself.

6. Conclusions

In the past few years orbital elements have been determined for binaries more than an order of magnitude younger than previously studied binaries of comparable mass (roughly 0.5 M_O - 2 M_O). The motivation for pushing to younger and younger ages is that *in situ* observations of very young binaries will provide insights into binary formation and evolution different from those derived from the boundary conditions provided by main-sequence binaries. This paper has focussed on the pre-main sequence eccentricity distribution in particular. While the number of orbital eccentricities determined for pre-main sequence binaries needs to be greatly increased, several results and conclusions are beginning to develop.

1) Orbital eccentricities from e=0 to e=0.8 are found in pre-main sequence binaries; substantial orbital eccentricity is present shortly after, and likely prior to, their unveiling at the stellar birthline. The pre-main sequence and main-sequence orbital eccentricity distributions are very similar for periods where the observations overlap (< 200^d), except for differences in observed circularization cutoff periods. Thus the eccentricity distribution found among main-sequence binaries is largely established by ages of $\approx 10^6$ yr, at least for binaries no longer having associated disks.

2) The pre-main sequence eccentricity distribution is clearly period dependent with several notable features. The four shortest period binaries (P < 4.3^d) all have circular orbits. There is an upper envelope on the eccentricity distribution between the circularization cutoff period and periods of several hundred days where the maximum eccentricity increases with increasing period. And there is a lack of near-circular (e < 0.1) orbits for periods above the circularization cutoff period.

3) The presently known pre-main sequence binaries set a circularization cutoff period of 4.3 days, although only two (eccentric) orbits are known with periods between 4.3 days and 10 days. The near-ZAMS binary EK Cep, with an eccentric orbit at a period of 4.4 days, alone contradicts the recent theoretical prediction of Zahn and Bouchet that pre-main sequence tidal circularization will be complete for binaries with periods of less than 7.3-8.5 days [ZaB89]. However this single case is not decisive.

4) At the shorter periods, pre-main sequence stars in binaries with eccentricities greater than the observed upper envelope would have frequent close encounters at periastron separations of less than a few stellar radii. Many dissipative processes, including tidal circularization, would be enhanced at such close separations, arguably causing such binaries to rapidly evolve in period and to lower eccentricity. As yet no specific mechanism has been sufficiently explored to quantitatively model the upper envelope.

5) We suggest that the lack of near-circular orbits among both main-sequence and pre-main sequence binaries for periods longer than the circularization cutoff periods may be the signature of disk excitation of eccentricity. Masses measured for disks around single pre-main sequence stars and in pre-main sequence wide binaries are theoretically sufficient to drive such evolution.This hypothesis implicitly presumes that disks of sufficient mass are present at least early in the lifetimes of all binaries.

6) However, the low-eccentricity orbit of GW Ori, a 242^d period binary showing many diagnostics for an associated massive disk, may pose a challenge to dynamical theories for binary-disk evolution. It also may indicate that disk structure in young binary environments does not follow our present simple characterizations (e.g., coplanar thin disks).

At this early stage of both observations and theory, perhaps it is still worth stressing that the eccentricity distribution of young binaries is likely the result of a complex interaction involving stellar structure, associated disks, stellar encounters and other processes. At present it is not clear that any of the key theoretical ingredients are sufficiently well understood: PMS stellar evolution in the presence of accretion, tidal circularization theory, binary-disk dynamics, multiple star formation, etc. Nonetheless, as these theories are further developed the pre-main sequence orbital eccentricity distribution will remain a key constraint.

The unpublished orbits presented in Table 1 are the result of collaborative efforts with D. Latham, L. Marschall, P. Myers and F. Walter. Many of the radial-velocity spectra were obtained by R.Davis, E. Horine, J. Peters and R.Stefanik. C. Clarke and S. Lubow provided many illuminating discussions at this meeting, and I would like to acknowledge C. Clark for raising the issue of periastron encounters. I thank W. Caplan, E. Jensen and S. Lubow for critical readings of the manuscript, and A. Duquennoy for electronic versions of the CORAVEL results. I also gratefully acknowledge the support of National Science Foundation Grant AST8814986, the Presidential Young Investigator program and the Wisconsin Alumni Research Foundation.

References

[AEF90] Adams, F.C., Emerson, J.P., Fuller, G. *Submillimeter photometry and disk masses of T Tauri disk system.* 1990, ApJ, 357, 606

[ALH89] Andersen, J., Lindgren, H., Hazen, M.L., Mayor, M. *The pre-main sequence binary system AK Scorpii.* 1989, AA, 219, 142

[ACL91] Artymowicz, P., Clarke, C.J., Lubow, S.H., Pringle, J.E. *The effect of an external disk on the orbital elements of a central binary.* 1991, ApJ, 370, L35

[BSC90] Beckwith, S.V.W., Sargent, A.I., Chini, R.S., Gusten, R. *A survey for circumstellar disks around young stellar objects.* 1990, AJ, 99, 924

[BeS92] Beckwith, S.V.W., Sargent, A.I. *The occurrence and properties of disks around young stars.* 1992, in Protostars and Planets III, eds. M. S. Matthews & E. Levy (University of Arizona Press: Tucson), in press

[Byr86] Byrne, P.B. *HDE 319139.* 1986, Irish AJ, 17, 294

[BRM92] Bodenheimer, P., Ruzmaikina, T., Mathieu, R.D. *Stellar multiple systems: Constraints on the mechanism of origin.* 1992, in Protostars and Planets III, eds. M. S. Matthews & E. Levy (University of Arizona Press: Tucson), in press

[ClP91] Clarke, C.J., Pringle, J.E. *The role of discs in the formation of binary and multiple star systems.* 1991, MNRAS, 249, 588

[DQT86] de la Reza, R., Quast, G., Torres, C.A.D., Mayor, M., Meylan, G., Llorente de Andres, F. *Simultaneous UV-optical observations of isolated T-Tauri stars: The V4046 Sgr. Case.* 1986, in New Insights in Astrophysics (ESA SP263), 107

[DuM91] Duquennoy, A., Mayor, M. *Multiplicity among solar-type stars in the solar neighborhood. II Distribution of the orbital elements in an unbiased sample.* 1991, AA, 248, 485

[DMM92] Duquennoy, A., Mayor, M. Mermilliod, J.-C. *Evolution of solar-mass binary orbital elements.* 1992, this volume

[Ghe92] Ghez, A. 1992, Ph.D. dissertation, California Institute of Technology

[GoM91] Goldman, I., Mazeh, T. *On the orbital circularization of close binaries.* 1991, ApJ, 276, 260

[JCM92] Jensen, E.L., Caplan, W., Mathieu, R.D., Lee, C.-W., Walter, F.M. 1992, in preparation

[JoB44] Joy, A.H., van Biesbrock, G. *Five new double stars among variables of the T-Tauri class.* 1944, PASP, 56, 123

[LHR92] Leinert, C., Haas, M., Richichi, A., Zinnecker, H., Mundt, R. *Lunar occultation and near-infrared speckle observations of DG Tau, FV Tau, FW Tau and GG Tau.* 1992, AA, in press

[LuA92] Lubow, S.H., Artymowicz, P. *Eccentricity evolution of a binary embedded in a disk.* 1992, this volume

[MaM88] Marschall, L.M., Mathieu, R.D. *Parenago 1540: A pre-main-sequence double-lined spectroscopic binary near the Orion Trapezium.* 1988, AJ, 96, 1956

[Mat92] Mathieu, R.D. *Disks in the pre-main sequence binary environment.* 1992, in IAU Symposium 151, Evolutionary Processes in Interacting Binary Stars, eds.Y. Kondo, R. Sistero and R. Polidan (Kluwer:Dordrecht), in press

[MAL91] Mathieu, R.D., Adams, F.C., Latham, D.W. *The T-Tauri spectroscopic binary GW Orionis.* 1991, AJ, 101, 2184

[MML92] Mathieu, R.D., Marschall, L.M., Latham, D.W. 1992, in preparation.

[MWM89] Mathieu, R.D., Walter, F.M., Myers, P.C. *The discovery of six pre-main-sequence spectroscopic binaries.* 1989, AJ, 98, 987

[MWM92] Mathieu, R.D., Walter, F.M., Myers, P.C. 1992, in preparation.

[MaM84] Mayor, M., Mermilliod, J.-C. *Orbit circularization time in binary stellar systems.* 1984, in IAU Symposium 105, Observational Tests of Stellar Evolution Theory, eds. A. Maeder and A. Renzini (Dordrecht, Reidel), 145

[Maz91] Mazeh, T. *Eccentric orbits in samples of circularized binary systems: The fingerprint of a third star.* 1991, AJ, 99, 675

[MWF83] Mundt, R., Walter, F.M., Feigelson, E.D., Finkenzeller, U., Herbig, G. Odell, A.P. *Observations of suspected low-mass post-T Tauri stars and their evolutionary status.* 1983, ApJ, 269, 229

[PaS90] Palla, F., Stahler, S.W. *The birthline for intermediate-mass stars.* 1990, ApJ, 360, L47

[Pop87] Popper, D.M. *A pre-main-sequence star in the detached binary EK Cephei.* 1987, ApJ, 313, L81

[RLN90] Reipurth, B., Lindgren, H., Nordstrom, B., Mayor, M. *Spectroscopic pre-main sequence binaries.* 1990, AA, 235, 197

[SCH92] Simon, M., Chen, W.P., Howell, R.R., Benson, J.A., Slowik, D. *Multiplicity among the young stars in Taurus*. 1992, ApJ, 384, 212

[SBW91] Skinner, S.L, Brown, A., Walter, F.M. *A search for evidence of cold dust Around Naked T Tauri Stars*. 1992, AJ, 102, 1742

[Tom83] Tomkin, J. *Secondaries of eclipsing binaries V. EK Cephei*. 1983, ApJ, 271, 717

[Zah89] Zahn, J.P. *Tidal evolution of close binary stars: I. Revisiting the theory of the equilibrium tide*. 1989, AA, 220, 112

[ZaB89] Zahn, J.P., Bouchet, L. *Tidal evolution of close binary stars: II. Orbital circularization of late-type binaries*. 1989, AA, 223, 112

[Zin90] Zinnecker, H. *Pre-main sequence binaries*. 1989, in Low Mass Star Formation and Pre-Main Sequence Objects, ed. B. Reipurth, (ESO:Munich), 447

A New Algorithm to Derive the Mass-Ratio Distribution of Spectroscopic Binaries

Tsevi Mazeh *   *Dorit Goldberg* †

Abstract

We present a new algorithm to derive the mass-ratio distribution of an observed sample of spectroscopic binaries. The algorithm replaces each binary of unknown inclination by an ensemble of virtual systems with a distribution of inclinations. This work shows that contrary to a widely held assumption the orientations of each virtual ensemble should *not* be distributed randomly in space. The algorithm comprises a few iterations to find the true mass-ratio distribution.

We also suggest a way to account for an observational selection effect associated with the inability to detect single-line spectroscopic binaries with low-amplitude radial-velocity modulations. We present several numerical simulations to demonstrate the advantage of the new algorithm over the classical method.

1.   Introduction

Large samples of spectroscopic binaries have been recently detected by precise systematic radial-velocity surveys (e.g. Latham *et al.* [LMC88]; Duquennoy, Mayor and Halbwachs [DMH91]). These samples can help to resolve the long standing controversy over the mass-ratio distribution of short-period binaries (e.g. Abt and Levy [AL76]; Halbwachs [Hal87]; Trimble [Tri90]). For example, we might be able to find out whether the distribution is different for long and short period binaries ([AL76]; Duquennoy and Mayor [DM91]). and whether the mass distributions of the primaries and the secondaries are similar to that of the single field stars ([AL76]; Tout [Tou91]). Unambigious answers to these questions can provide important clues to the formation of close binary systems (e.g. Bodenheimer, Ruzmaikina and Mathieu [BRM92]).

The derivation of the mass-ratio distribution of a given sample of spectroscopic binaries is confronted by two well known problems [Ait35]. First, the mass ratio of a

---

*School of Physics and Astronomy, Tel Aviv University, Israel, and Harvard-Smithsonian Center for Astrophysics, USA

†School of Physics and Astronomy, Tel Aviv University, Israel

specific system can be deduced only for double-line spectroscopic binaries. For the single-line binaries the mass ratio is unknown; only the mass function can be derived directly from the observations. For a binary system with a primary mass M_1 and a secondary mass M_2, the mass function is

$$f(M_2) = M_1 \frac{q^3}{(1+q)^2} \sin^3 i \quad , \tag{1}$$

where q is the mass ratio ($= M_2/M_1$) and i is the inclination of the binary orbit relative to our line of sight. Since the inclination is not known, the mass ratio can *not* be derived, even when the primary mass can be estimated from its spectral type.

Second, observational limits affect the detected binary sample. For example, systems with small radial-velocity amplitudes are not easily observed. so that binaries with small mass-ratios are sometimes not detected.

The problem of unknown inclinations was already addressed back in the 1920's (e.g. Aitken [Ait35]), and since the 1970's by many workers (see [Hal87], [Tou89] and [Tri90] for reviews). One statistical approach was to assign the expected value of $\sin^3 i$ to *all* single-line spectroscopic binaries of the sample. Aitken quoted Campbell and Schlesinger, who had suggested to use an average value of 0.589 for $\sin^3 i$ in an ideal unbiased sample of binaries. They derived this value by averaging $\sin^3 i$ over all possible angles. (For real samples, Campbell and Schlesinger preferred a value of 0.679, obtained by assuming a detection probability that depends on the binary inclination and is proportional to $\sin i$.) The philosophy behind this method is clear: for a large enough sample, the differences between the assigned values of $\sin i$ and the true values are averaged to zero.

A different statistical approach was suggested by Jaschek and Ferrer [JF72], and was adopted by Halbwachs [Hal87] and Trimble [Tri90]. This indirect approach starts with an assumed function for the mass-ratio distribution of the sample. A prediction for a distribution of the *mass-function* divided by the primary mass is then derived, and compared with the actual observed one. One then finds, from this comparison, the best parameter(s) of the assumed function. This model-fitting approach enables us to find only the best parameters of an *assumed* function, but can not be used to find new, possibly better, functions to describe the mass-ratio distribution. We suggest therefore that the preferred approach is a direct method. which does not assume any function, but derives the true distribution directly from the data. Such a direct method is the one of Campbell and Schlesinger (hereafter CS). However, this method fails to reconstruct the true mass-ratio distribution. This is because one of its basic assumptions is incorrect. In this work we demonstrate the failure of the CS method, by using some simulated samples, and point out the specific reasons for its failure.

This work suggests a new iterative statistical algorithm. which replaces each binary with an ensemble of virtual systems of non-random orientations. In order to present the new algorithm, we first ignore the observational selection effects, and use the new method to analyze idealized samples of binaries. Numerical simulations show the success of the new method to reconstruct the right mass-ratio distribution. We further suggest how to correct for an observational selection effect associated with the inability to detect spectroscopic binaries with low-amplitude modulations. We present one example in which this observational effect was simulated. The example shows that the proposed procedure reproduces the correct mass-ratio distributions.

2.   The failure of the Campbell and Schlesinger direct method

In order to demonstrate the failure of the CS method, we applied it to a *simulated* sample of 2000 binaries, with a uniform mass-ratio distribution and with random orientations. We set the period distribution of the sample to be uniform in $\log P$, between 1 and 1000 days, and the primary mass to be 1 $M_\odot$. We then calculated the mass function for each binary by using Eq. (1). The simulated sample was then analyzed, using the only information available for each binary in real samples — the primary mass and the mass function, ignoring the information about the mass ratio of each binary.

The results of this exercise are presented in Figure 1a. Clearly, the CS method is drastically biased toward the low end of the mass-ratio spectrum. In Figure 1b we present a similar simulation with mass-ratio distribution which rises toward unity. The intrinsic bias of the CS method is very prominent again. Many of our simulations yielded similar results.

To understand the reasons for the failure of the CS method, we must first clarify its basic assumptions. Explicitly, the method assumes that the orbital planes of the sample are randomly oriented. However, this assumption is not enough. Actually, the CS method implicitly assumes random orientations within certain subsets of the sample — subsets with a particular value of the mass-function. Otherwise, it would not be possible to assign the same value of 0.589 to every binary of the sample.

This further assumption is not correct, because the mass function is not an independent variable of the sample. Our basic assumption is that the independent variables are the mass-ratio, the inclination, and possibly the periods of the binaries. The mass function depends on these variables through Eq.(1), and therefore is a dependent variable.

To demonstrate this point we plot in Figure 2 some constant mass-function contours on the $q - \sin i$ plane, for primary mass of $1 M_\odot$. The two axes represent independent variables, while clearly the contours do not. As can be seen from Figure 2, once we choose the mass-function between some values f and $f + df$, the two variables i and q are not independent anymore. Moreover, the probability of a system with a mass-function between those values to have a certain inclination i depends on the

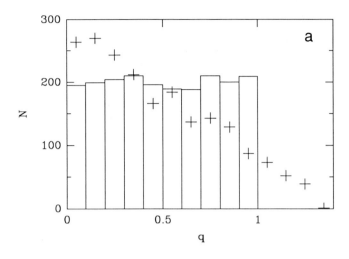

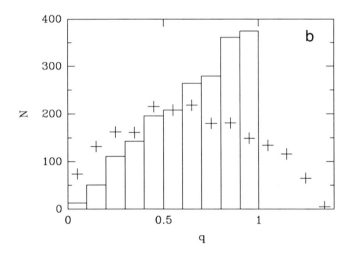

Figure 1.   Numerical simulations to test the classical Campbell and Schlesinger method. The histogram shows the true distribution of the simulated sample. The pluses represent the results of the method.
a. A simulation with a uniform distribution.
b. A simulation with a monotonously increasing distribution of $N(q) = 2q$.

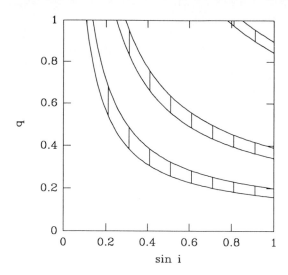

Figure 2. 'Stripes' of constant mass function in the mass-ratio – $\sin i$ plane. The three 'stripes' are confined by the values (0.001, 0.002), (0.01, 0.015) and (0.15, 0.18) respectively. Lines parallel to the q axis cover the stripes. Their varying length reflects the fact that the partial derivative of f with respect to the mass ratio varies as a function of $\sin i$.

mass-ratio distribution as well as on the $\sin i$ distribution. To illustrate this point consider an extreme situation where all the binaries of the sample have only one q value – q_0. In such a case the probability of a system with a mass-function f_0 to have inclination i is different from zero only for i_0, which solves Eq. (1) for given q_0 and f_0. Thus the inclination distribution is not random, and depends on the specific value of f and on the q distribution.

The assumption that the orientations of the subsample are randomly distributed introduces three sources of error to the analysis:

1. For samples that include only systems with mass-ratio smaller then unity, any given subsample of mass-function f does not include inclinations smaller than a certain limit (see figure 2). The CS method averages over all angles, and does not exclude inclinations below the lower limit. Thus, it underestimates $\sin^3 i$, and in turn overestimates q for most binaries.

2. In order to average $\sin^3 i$ over all possible angles, we need to know the mass-ratio distribution. This is, however, precisely the distribution that we try to extract from the sample! The CS method assumes a uniform mass-ratio distribution while averaging over the inclination angles. This is at best a zero-order approximation, which can induce serious bias. Clearly, to solve this problem we need some iterative approach.

3. For every subsample with a mass-function between f and $f + df$ there is an allowed stripe in the $\sin i - q$ plane. (Three such stripes are plotted in Figure 2.) When averaging, one should integrate only over the allowed stripe. Figure 2 shows that the width of these stripes, when measured parallel to the q axis, varies as a function of $\sin i$. This results from the fact that the partial derivative of the mass function with regard to q varies as a function of $\sin i$. The variation of the finite width introduces another deviation from the random distribution.

Replacing $\sin^3 i$ by its expectation value introduces another source of error. As pointed out by Halbwachs [Hal87] in his seminal work, the specific shape of the distribution of $\sin^3 i$ is heavily weighted toward the two extreme points, where $\sin^3 i$ equals zero or unity. (See Figure 2 of [Hal87] and the discussion there.) The average value of $\sin^3 i$, 0.589, used by the CS method, is, therefore, far from being the most probable value. This feature led Halbwachs to reject the CS method as an appropriate approach to the problem.

The CS method, despite all its drawbacks, was a very important step in the way to derive the true mass-ratio distribution. We have to bear in mind that the method was conceived and first used when computing machines were not available. The beauty of this zero-order approximation relies on the fact that no computing is needed, and the averaging can be done analytically. The next section outlines our modified approach which takes into account the errors mentioned above. The new algorithm takes some amount of numerical computing, a feature which should not be considered disadvantageous in these days, contrary to the situation in the good old days of Aitken.

3.   The proposed algorithm

The proposed algorithm considers each observed binary as drawn from a large subset of binaries with different inclinations, but with the same period, primary mass, and mass-function. To take into account the special shape of the inclination distribution, the algorithm replaces each binary with an ensemble of N virtual systems, which mimics the parent subset of the binary. For normalization purposes, each of the virtual systems represents $1/N$ binaries. For each virtual system we know the mass function, the primary mass and the inclination, and therefore can solve its mass ratio. The mass-ratio histogram of the ensembles of all binaries included in the sample represents our best estimation of the mass-ratio distribution of the observed sample.

Three main features characterize each ensemble:

1. The algorithm finds for each binary a lower limit for the possible inclination, i_{min}.

Inserting $q = 1$ into Eq. (1) we get:

$$\sin^3 i_{min} = \frac{4f(M_2)}{M_1} \quad . \tag{2}$$

The upper limit of the inclinations is, of course, $\sin i = 1$.

2. The algorithm performs few iterations in order to account for the dependence of the inclination distribution on the unknown mass-ratio distribution. To begin, we assume a uniform distribution of q, and assign each virtual ensemble the resulting inclination distribution. We then derive the q distribution of the whole sample. This new distribution is the first-order approximation derived by the algorithm. It is then used as the input for calculating the second-order iteration, and so on. This process is continued till the n-th order approximation is statistically indistinguishable from the (n-1)-th one.

3. The variation of the partial derivative of the mass function with regard to q can be represented by

$$A(q) = \frac{(1+q)q}{3+q} \quad . \tag{3}$$

To account for this effect, the algorithm multiplies each distribution by this factor.

To test the proposed new algorithm we applied it to the same simulated samples that were used to demonstrate the drawbacks of the CS method. The results are presented in Figure 3. A comparison of Figures 1 and 3 shows the advantage of the proposed algorithm over the CS method. Numerous tests with different samples yielded the same conclusion.

The algorithm requires a few further subtleties. We have found that the iterations converge only if the input mass-ratio distribution for each iteration is continuous. Because the resulting histogram of each iteration is not continuous, we interpolated between the bins to get the input distribution for the next iteration. Another subtle point is handling the double-line spectroscopic binaries (SB2). The mass-ratios of the SB2's are known, and therefore there is no need to replace these binaries with ensembles of virtual systems. Instead, we first find the mass-ratio distribution of the single-line binaries using the new algorithm, and then add the SB2's. We have run numerous simulations to test the new algorithm, some of them included also SB2 systems. All the simulations converged, all to the right solution, and with no dependence on the first guess.

We could not resist the temptation to try the new algorithm on real binary samples. Obviously, the real samples are smaller than the simulated ones, and are biased by selection and detection effects. We applied the algorithm to two samples. The first sample is that of Duquennoy and Mayor [DM91], which includes all the 37 spectroscopic binaries they found, out of the nearby G stars. The other sample consists of 57 halo spectroscopic binaries, with metallicity less than $[m/H] = -1.2$ (Latham *et*

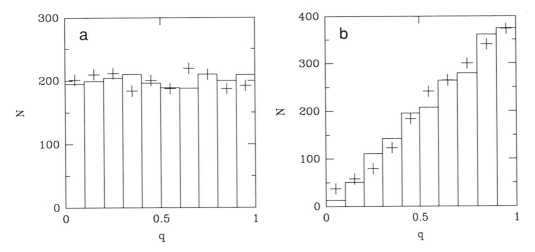

Figure 3. Numerical simulations to test the proposed new algorithm. The histogram shows the true distribution of the simulated sample. The pluses represent the results of the algorithm. The figure should be compared with Figure 1.
a. A simulation with a uniform distribution.
b. A simulation with a monotonously increasing distribution of $N(q) = 2q$.

al. [LTC92]), out of the Carney-Latham large survey (Carney and Latham [CaL87]). The results are shown in Figure 4. Both samples indicate an increasing distribution toward unity. We would like to empasize, however, that the two distributions were derived without any correction for the selection effects of the two samples. Therefore, the results should *not* be considered as the true mass-ratio distributions of spectroscopic binaries, but only as experiments to get the *observed* distributions.

4. Compensating for undetected binaries

We turn now to discuss one of the observational selection effects found in samples of spectroscopic binaries, the one associated with the inability to detect binaries with low-amplitude radial-velocity modulations. We suggest a new procedure to compensate for this effect.

Specifically, we consider a radial-velocity survey of a well-defined preselected sample of stars, of which any binary with radial-velocity amplitude larger than K_{min} is detected. This selection effect applies, to first-order approximation, to some of the large recent surveys ([LMC88]; [LTC92]; [DM90], [DM91], [DMM92]). Note that this observational selection effect is very different from the one assumed by Trimble [Tri74], [Tri90]), Staniucha [Sta79]) and Halbwachs [Hal87], because the recent samples are very different from samples included in any catalog of known spectroscopic binaries (e.g. Batten, Fletcher and MacCarthy [BFC89]).

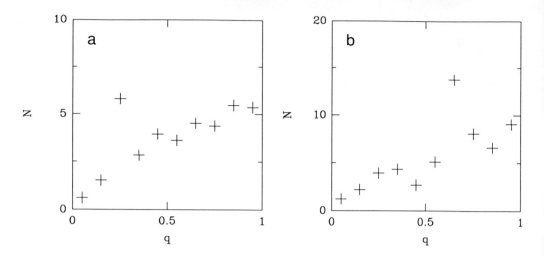

Figure 4. The results of the new algorithm applied to real binary samples, without correcting for any observational selection effect.
a. The sample of Duquennoy and Mayor (1991) — 37 G-dwarf binaries.
b. The sample of Latham *et al.* (1992) — 57 halo binaries.

The radial-velocity amplitude of the primary, K_1, can be expressed as:

$$K_1 = 212.9 \, P^{-1/3} \, M_1^{1/3} \, \frac{q}{(1+q)^{2/3}} \, \sin i \quad (\text{km/s}) \quad , \tag{4}$$

where P is the orbital period in days and M_1 is in solar mass units. We consider here, for simplicity, only circular orbits. Consider now a subsample of binaries, all with the same binary period, mass ratio, and primary mass, and with a given distribution of inclinations. Out of this subsample, all systems with an inclination smaller than some minimal inclination, i_0, are *not* detected. One gets from Eq. (4) that:

$$\sin i_0 = 4.697 \times 10^{-3} \, K_{min} \, P^{1/3} \, M_1^{-1/3} \frac{(1+q)^{2/3}}{q} \quad . \tag{5}$$

We therefore define a detection function, $D_{K_{min}}(P, q, M_1)$, which is the fraction of detected binaries out of all binaries with the same P, q and M_1:

$$D_{K_{min}}(P, q, M_1) = \int_{\sin i_0}^{1} \Psi(i) \, d(i) = \sqrt{1 - \sin^2 i_0} \quad , \tag{6}$$

where Ψ is the random distribution of the angles. Putting Eq. (5) into Eq. (6) gives for circular orbit:

$$D_{K_{min}}(P, q, M_1) = \sqrt{1 - \left[4.7 \times 10^{-3} \, P^{1/3} \, K_{min} \, M_1^{-1/3} \frac{(1+q)^{2/3}}{q}\right]^2} \quad . \tag{7}$$

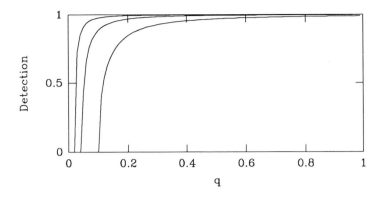

Figure 5. The detection function as a function of the binary mass-ratio, for different values of the period P (10, 100, and 1000 d). The detection threshold was taken as $K_{min} = 2$ km/s.

The dependence of D on P and q, as expressed in Eq. (7), is plotted in Figure 5, for $K_{min} = 2$ km/s, and for $M_1 = 1M_\odot$. The function is close to unity for most values of P and q, except for low q values, where the function is very steep. The low values of D for small q indicate that any sample of spectroscopic binaries is vulnerable to large statistical errors in this range of the mass-ratio domain. We therefore regard the derived mass-ratio distribution for small q values as highly uncertain, unless K_{min} is very small. The detection function enables us to assign each binary a correction factor,

$$C(K_{min}, P, q, M_1) = \frac{1}{D} \ ,$$ (9)

if we know its period, primary mass, and its mass ratio in particular. To compensate for the undetected binaries, we consider each binary as representing C number of systems with the same parameters. Unfortunately the mass ratio of each binary is not known in real samples, and the correction factor can not be applied. The correction factor procedure can work, however, within the proposed new algorithm. Each virtual system is assigned a mass ratio, and therefore its correction factor can be derived and used.

The advantage of the procedure outlined here is its capability to apply the correction for the undetected binaries to each virtual system independently. However, for small enough mass ratios, even a single virtual system is not assigned; the specific values depend on the range of observable periods of the survey and on K_{min}. In such a case, the individual correction procedure can not be applied, and the number of systems

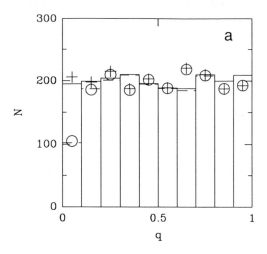

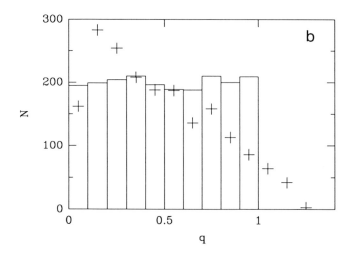

Figure 6. Numerical simulations to compare the two methods, when applied to a sample with a detection threshold, K_{min}, of 2 km/s. The continuous-line histogram represents the true distribution of the original simulated sample, including all binaries. The broken-line histogram shows the distribution of the 'detected' binaries.

a. The results of the new algorithm — the circles represent the derived distribution of the detected binaries, while the pluses show the obtained total distribution.

b. The results of the old CS method, assuming a detection probability proportional to $\sin i$.

with small mass ratios has to be extrapolated. The extrapolation must assume some period distribution, the natural choice being the observed period distribution of the whole sample. This correction introduces, again, a large uncertainty to the low end of the derived mass-ratio distribution. In the recent surveys of the CORAVEL and the CfA stellar speedometers this problem usually appears only for mass ratios smaller than 0.1.

To demonstrate the effectiveness of the new procedure we, again, used a simulated sample of binaries. We used the same sample with uniform mass-ratio distribution. Out of this sample we generated a subsample of the 'detected' binaries. in which we included only binaries with an amplitude larger than a threshold detection of $K_{min} = 2$ km/s. We first applied our modified iterative algorithm, to find the mass-ratio distribution of the *detected* sample. We then used the correction factor to find out the true distribution of the whole sample. The results are depicted in Figure 6, together with the CS method result, where we assigned to each binary a value of 0.679 for $\sin^3 i$, as suggested by Campbell and Schlesinger for real samples. The Figure suggests that the combination of the iterative proposed algorithm with the individual correction factor procedure is useful in some cases.

5.   Conclusion

We have proposed a modified direct approach to derive the mass-ratio distribution of an observed sample of spectroscopic binaries. The effectiveness of the new algorithm was demonstrated with several illustrative numerical simulations. in which we were able to reconstruct the initial distribution of idealized samples of binaries. We have outlined a new procedure to correct for one important observational effect associated with the detection limit of the radial-velocity amplitude. However, a few more observational selection effects should be accounted for when deriving the true distribution of any sample. We therefore defer any analysis of the very recent samples to subsequent work, where the algorithm will be applied and the observational selection effects will be addressed thoroughly. Hopefully, the new data will enable us to determine soon the true shape of the mass-ratio distribution of stellar binaries.

Acknowledgements. It is a pleasure to express our deep gratitude to A. Sternberg for many enlightening discussions. We thank A. Duquennoy, I. Goldman, J.-L. Halbwachs, D. Latham, M. Mayor and G. Torres for many useful discussions which we greatly enjoyed. This work was supported by the U.S.-Israel Binational Science Foundation Grant No. 90-00357.

References

[AL76] Abt, H. & Levy, S.G. *1976, ApJS, 30, 273*

[Ait35] Aitken, R.G. *1935 in The Binary Stars (McGraw-Hill), 218*

[BFC89] Batten, A.H., Fletcher, J.M. & MacCarthy, D.G. *1989, Publs Dom. astrophys. Obs., 17, 1*

[BRM92] Bodenheimer, P., Ruzmaikina, T. & Mathieu, R.D. *1992, in Protostars and Planets III, eds. M.S. Mathews & E. Levy (Tucson: University of Arizona Press).*

[CaL87] Carney, B.W. & Latham, D.W. *1987, AJ. 93, 116*

[DM90] Duquennoy A. & Mayor, M. *1990, in New Windows to the Universe, Proc. XI-th European Regional Astronomy Meeting of the IAU, eds. F. Sanches & M. Vasquez (Cambridge: University Press), 253*

[DM91] Duquennoy, A. & Mayor, M. *1991, A&A, 248, 485*

[DMH91] Duquennoy, A., Mayor, M. & Halbwachs, J.L. *1991, A&AS, 88, 281*

[DMM92] Duquennoy, A., Mayor, M. & Mermilliod, J.C. *1992, in Binaries as tracers of stellar formation, Ed. A. Duquennoy & M. Mayor, (Cambridge: Univ. Press)*

[Hal87] Halbwachs, J.L. *1987, A&A, 183, 234*

[JF72] Jaschek, C. & Ferrer, O. *1972, PASP, 84, 292*

[LMC88] Latham, D.W., Mazeh, T., Carney, B.W., McCrosky, R.E., Stefanik, R.P. & Davis, R.J. *1988, AJ, 96, 567*

[LTC92] Latham, D.W., Torres, G., Carney, B.W. & Mazeh, T. *1992, in Binaries as tracers of stellar formation, Ed. A. Duquennoy & M. Mayor, (Cambridge: Univ. Press)*

[Sta79] Staniucha, M. *1979, Acta Astr., 29, 587*

[Tou89] Tout, C.A. *1989, Ph.D. thesis, Cambridge Univ.*

[Tou91] Tout, C.A. *1991, MNRAS, 250, 701*

[Tri74] Trimble, V. *1974, AJ, 79, 967*

[Tri90] Trimble, V. *1990, MNRAS, 242, 79*

Distribution of Orbital Elements for Red Giant Spectroscopic Binaries in Open Clusters

Jean-Claude Mermilliod [*]   *Michel Mayor* [†]

Abstract

Systematic Coravel observations of 1048 red giant stars in 177 open clusters resulted in the discovery of 187 spectroscopic binaries. The frequency of spectroscopic binaries with periods less than about 10^4 days among open cluster red giants is 23% (144/638). The preliminary analysis of a sample of 88 orbits for red giant members shows that (1) in the mass range $2 < M_1 < 3\ M_\odot$, the distribution of mass ratio is flat, (2) the mass-function distribution probably depends on the primary mass.

1. Introduction

There are many reasons for observing the radial velocity of red giants in open clusters. A very basic one is simply that red giants in open clusters have never been subjected to any systematic study. At the epoch (1977) we initiated this observing program, only two orbits were known, one for Tr 91 in Coma Berenices [Vin40] and one for vB 41 in the Hyades by Griffin and Gunn [GrG77]. The situation has considerably changed in the past years. About 190 spectroscopic binaries with a red giant primary have been discovered and observed continuously. Orbital elements for 22 red giant binaries have already been published by Mermilliod and Mayor [MeM89], [MeM90] and Mermilliod et al [MMA89]. Orbits for 8 red giants in M67 have been published by Mathieu et al [MLG90] and for 2 in M11 (NGC 6705) by Lee et al [LML89].

The determination of the distributions of orbital elements for spectroscopic binaries with a red giant primary is another important reason for observing these stars, because we know at the same time their mass and age, thanks to their membership of open clusters. Since the usual precision obtained with photoelectric scanners is equal to or better than 1.0 km/s, we can discover and observe binaries with long periods and small amplitudes (K $\geq$ 2 km/s). Such binaries were practically unobservable

[*]Institut d'Astronomie de l'Université de Lausanne, 51 chemin des Maillettes, CH 1290 Chavannes-des-Bois, Switzerland

[†]Observatoire de Genève, 51 chemin des Maillettes, CH 1290 Sauverny, Switzerland

among main sequence B- and A-type stars, which are the progenitors of most red giants we observe. Thus the combined sample of main sequence and red giant stars should produce a rather complete distribution of orbital elements.

However it is not possible to observe the original distribution because: (a) main sequence binaries with short periods (a few times 10 days) probably never reach the red giant stage since the evolution of the primary star leads to Roche lobe overflow, (b) binaries with a red giant primary of intermediate mass and an orbital period in the range $40 < P < 120\text{-}150^d$ are circularized due to tidal interaction between the two components, (c) it is not sure that the binaries we observe are all primordial ones. It may happen that the secondary is a white dwarf and that the orbital elements have been modified at any stage of the evolution of the former primary, (d) a number of primordial binaries are now under the form of a main sequence star and a white dwarf, after the former primary completed its evolution. These questions complicate the analysis of orbital element distributions and bear some importance on the possible tests of the various stellar formation and evolution models.

Radial velocity observations of red giant stars in open clusters are not only useful to investigate the binary properties, but also a very powerful method to determine their membership, provided there are at least two or three red giant members in a cluster. The knowledge of membership and binarity has allowed to determine empirical isochrones in the red giant region of the colour-magnitude diagram ([MeM89], [MeM90]) in clusters of the Hyades generation ($8.80 < \log t < 9.10$). We will also use these observations to study the dynamics of stellar systems. There are probably a few rich clusters for which the internal velocity dispersion is significantly larger than the observational errors.

To date, 144 spectroscopic binaries have been discovered with our Coravel observations among confirmed members, and 78 (54%) orbits have been determined. The present study will include the data published by other observers, in Coma Berenices [Vin40], the Hyades [GrG77], M67 [MLG90] (for which Coravel observations were obtained on request from Dr R. Griffin), and M11 [LML89]. The total sample contains then 663 red giants, 152 binaries and 88 orbits. Many binaries have long periods, often more than 4000 to 5000 days, and sometimes not even a single cycle has yet been covered. The diagrams presented here will therefore reflect the bias and gaps in the observations of binaries with periods larger than 2500^d.

2.   Sample and Observations

The observations were started in 1977 at the Geneva station of the Haute Provence Observatory (France) on the 1m Swiss telescope and in 1983 with the southern Coravel scanner installed on the 1.54m Danish telescope at the ESO Observatory (La Silla, Chile). The observational details have been described by Mermilliod and Mayor [MeM89]. The error on one observation is usually less than 1 km/s, and fre-

Table 1. List of observed clusters with at least 9 giant members

(1)	(2)	(3)	(4)	(5)	(6)	(7)	(8)	(9)	(10)	(11)	(12)
NGC 752	30	13		17	3		2	13	77-88		OHP
NGC 2099	55	21	7	34	5	2	5	4	81-88		OHP
NGC 2354	24	11	1	13				2	89-90		ESO
NGC 2360	24	5	1	19	4	1	3	4	83-89	*	ESO
NGC 2423	12	2		10	1	2		5	80-88	*	OHP
Mel 71	22	6		16	5	1	3	3	87-92		ESO
NGC 2447	13			13	3	1	2	4	83-92	*	ESO
NGC 2477	86	7	1	79	18	8	9	4	83-91		ESO
NGC 2539	13	2		11	5		5	6	78-89	*	OHP
IC 2714	14	5	1	9				3	84-90		ESO
NGC 3680	14	4	2	10	1		1	5	83-91		ESO
NGC 3960	13	3	1	10	4	2		3	84-90		ESO
NGC 4349	13	2		11	4		2	4	83-91		ESO
NGC 5822	28	7		21	8	2	6	4	83-90	*	ESO
NGC 6067	14	1		12	2			4	83-91	*	ESO
NGC 6134	23	5	1	18	6	2	2	3	85-90	*	ESO
NGC 6259	13	3	1	12				3	83-87		ESO
IC 4651	16			16	6		4	4	83-91		ESO
IC 4756	19	2		17	3		2	9	78-89	*	OHP
NGC 6940	24	3		20	6		6	6	78-89	*	OHP

quently around 0.5 km/s. The condition for inclusion of a cluster in the observing program is that all, or most, giants are brighter than the commonly used observing limit (B = 12.5 at OHP, and B = 14.5 at La Silla) so that we define an unbiased sample. In each cluster we have observed all stars which were considered as possible members from their position in the colour-magnitude diagram, especially those falling within the Hertzsprung gap.

The total sample contains to date 1048 stars in 177 open clusters. Although we have observed all possible stars, this figure represents only one third of the total number (3000) of red giants found in the field of open clusters. Table 1 gives the list of the observed clusters containing at least 9 red giant members. It gives successively: (1) the cluster identification, (2) the total number of stars observed, (3)-(4) information on non-members (number of non-members found and SB detected), (5)-(8) information on members (number of members, SBs, suspected binaries, orbits obtained), (9) number of observations for constant stars, (10) time span of the observations and (12) place of observations. An "*" in column (11) indicates that the study of the cluster has been published ([MMB87], [MeM89], [MeM90]). The analysis of NGC

6134 has been performed by Claria and Mermilliod [ClM92] and that for the three clusters NGC 752, 3680 and IC 4651 is being prepared for publication.

3.   Discussion

3.1.   The spectroscopic binary frequency

As time elapsed, the number of spectroscopic binaries discovered has grown. One of the longest waited for binaries, first suspected because its velocity was 3 km/s off the mean cluster velocity, proved to be finally variable after ten years of monitoring. It is therefore not surprising that our first estimate of the binary percentage, namely 18% [MeM89] appears now too small. The present value (144/638: 22.6% $\pm$ 1.9%) refers to spectroscopic binaries with periods less than $10^4 d$ and is not sensitive to the way the sample is formed, all member stars, clusters with more than 3 members, or with more than 10 members. The rejection of NGC 2477 (79 red giants) does not change this value either. The binary rate does not include the suspected binaries. The binary frequency for the field giant (non-member) stars deduced from our sample (see table 2) seems smaller: 16% $\pm$ 2%. Harris and McClure [HaM83] obtained a frequency of spectroscopic binaries, after three years of observations, of 15% to 20% in a sample of 40 K giant stars selected at random. However, due to the limited size of the samples, the observed differences are probably not significant.

A summary of the statistics based on the Coravel observations is given in Table 2. The sum of the number of confirmed members and non-members is not equal to the number of observed stars, because the membership is not defined when we have one red giant only in the cluster and no radial velocity for any main sequence star. Furthermore, only clusters with at least three observations per star have been taken into account in this statistics. The frequency obtained for the cluster red giants is very close to that deduced by Duquennoy and Mayor [DuM91] for the nearby G dwarfs with periods less than $10^4 d$, namely 21% $\pm$ 4% (34/164). However, if one considers only the interval $40 < P < 10^4 d$, the frequency of spectroscopic binaries among red giants (22.6%) is larger than that for the G dwarfs (13.4% $\pm$ 2.9%).

To analyse the dependence of the number of binaries in function of the cluster richness, estimated by the number of red giants, we have plotted a diagram with the number of spectroscopic binaries versus the number of red giant members (Figure 1). Although there are a few exceptions, the number of binaries increases with the number of red giants. The global slope of the relation (0.23) is shown by the long-dashed line. The error envelope, corresponding to the 1 σ limit, is indicated by the short-dashed lines. The points representing clusters containing up to 20 stars form a scattered sequence, but most observed points are located within the band defined by the error limits.

A few points are worth further comments. The richest cluster we have observed is NGC 2477 (79 gK; 18 SBs, 8SBs?). Its own binary frequency is 18/79 = 0.23. It falls

Table 2. Statistics over the Coravel sample

Red Giants observed	:	1048
Number of clusters	:	177
Member Stars	:	638
Spectroscopic Binaries	:	144
Suspected Binaries	:	30
Orbits obtained	:	78
Binary Rate (members)	:	0.23
Non Members	:	267
Spectroscopic Binaries	:	43
Binary Rate (non members)	:	0.16

exactly on the mean relation. The overall binary rate however does not depend on the value of NGC 2477. Eight stars are considered as possible binaries and half of them have a good probability of being binaries. The expected amplitudes are small and the periods probably long, so that it will require more time and observations to definitely settle their binary nature.

M67 (37 gK; 7 SBs; 19%) also falls within the relation. Only red giants in the Fagerholm area have been taken into account. The percentage will probably become lower if giants from the cluster corona are also counted.

NGC 2099, M37 (34 gK; 5 SBs; 15%) is also within the 1 σ limit. In fact seven more spectroscopic binaries were discovered, but all turned out to be non-members once the systemic velocities were obtained from the orbits.

The lowest point is NGC 6705, M11 (31 gK, 2 SBs; 6.5%). The probability of observing only two binaries among 31 stars is only 1%. This cluster very probably presents an anomalously small number of binaries. From the data published by Mathieu et al [MLG86] not more than one or two stars can be suspected of being low amplitude binaries.

In this context, the results for NGC 188 and 7789 based on the observations made at Victoria are very interesting. Harris and McClure [HaM85] announced in NGC 188, 33 members and 6 SBs, with 6 suspected variables. McClure (1992) [McC92] obtains 6 SBs (+ 7 SBs?) among 55 members in NGC 188, corresponding to 11% and 15 SBs (+ 8 SBs?) among 77 members in NGC 7789, giving a proportion of 19%. The

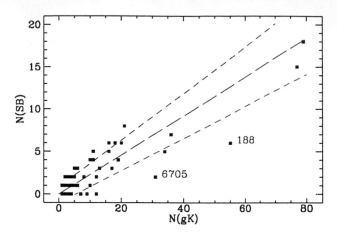

Figure 1. Number of spectroscopic binaries in function of the number of red giants. The long-dashed line represents the mean slope (0.23) and the short-dashed lines, the 1 σ limits.

position of NGC 7789 agrees quite well with the mean relation, while NGC 188 is rather low. It is striking that the observations of 22 more stars in NGC 188 between 1985 and 1992 did not change the number of spectroscopic binaries.

The reason for the binary deficiency in NGC 6705 is not known and probably not related to age or density, since NGC 2477 is only slightly older, much richer, and has nevertheless a normal binary frequency. One reason could be related to the completeness of the samples. If mass segregation is present, which has been shown by Mathieu [Mat84] for NGC 6705 (M11), one expects to find the binaries closer to the cluster center. Accordingly, the binary rate may depend on the size of the field surveyed. This argument is certainly not relevant for this cluster which has been quite extensively observed. Furthermore it cannot explain either the differences observed between NGC 188 and M11 on the one hand, and M37, M67, NGC 2477 and 7789 on the other. The observations of these clusters are the most complete ones and the fields have been surveyed to large distances of the cluster centers.

NGC 6259 presents also a slight anomaly since it contains 12 giant members, while no spectroscopic binaries were found. However, 7 more red giants found by Anthony-Twarog et al [APT89] and located outside the cluster nucleus have to be observed.

This last example points towards a major problem in this survey: what is the completeness of the detection of red giants at large distances (R > 1 cluster radius) from the cluster center? A search made at the telescope in NGC 6134 and IC 4651, by picking with Coravel stars which had about the right magnitude, as seen on the TV screen, resulted in the discovery of three more stars in the outer part of both clusters. Their radial velocities are very close to the cluster mean velocity and their membership is quite certain. The photometric observations obtained in the UBV system for NGC 6134 [ClM92] also support the membership. Observations in the Geneva system have been obtained for the new probable members in IC 4651, but have not been reduced yet.

A similar search in NGC 4349 based on quick photometric observations obtained with the Swiss telescope at La Silla, yielded again three additional member stars in the outer region of this cluster. This means that the real extension of several clusters is larger than thought from visual inspection of photographs or, at least, larger than the surface area investigated. Two more examples of outer lying red giants are given by NGC 2516 and 5662. In the former cluster, Dachs and Kabus [DaK89] found a new red giant (HD 64320) located at 1.2 deg from the cluster center, i.e. at 2.5 cluster radius. The radial velocities obtained with Coravel in 1990 and 1991 are very close to the mean value defined by the three other giants. In the latter cluster, NGC 5662, Claria et al [CLB91] have detected a probable red giant member at a distance of 30 arcmin from the center, while the cluster radius is about 15 arcmin. Again, two Coravel observations made in 1991 gave exactly the same velocity as that obtained for the star Haug 49. Conversely, in NGC 2360 [MeM89] and NGC 2423 [MeM90] a red giant has been discovered in the central crowded image not measured photometrically. Both red giants are spectroscopic binaries.

These examples are quite worrying. This could mean that many investigations were made over cluster surface areas that are too small in comparison with the real cluster size. This problem is also becoming important nowadays with CCD observations. In several cases, only one central field was taken, although clusters were obviously larger than the size of the CCD field. This will of course contribute to enhance this problem. Since open clusters have generally a small number of stars as compared to globular clusters, it is really important to know as many members as possible. This is especially true for the red giant and upper main sequence stars, because the ratio of their respective number is one of the main observational results to which predictions of stellar evolutionary models can be compared.

Table 3 proposes a list of old open clusters which are interesting due to the number of red giants they contain. Unfortunately, none of them can be observed completely with the Coravel scanners. It would however be important to enlarge the sample of old clusters containing numerous red giants systematically studied for radial velocity so as to investigate the possibility of a difference in binary rate from cluster to cluster, the binary segregation at different ages, and cluster richness.

3.2.   Distribution of orbital elements

The present sample (88 stars) represents about one half of the total number of known spectroscopic binaries in the 180 clusters systematically investigated for radial velocity with the Coravel scanners or other instruments. The distributions presented here are of course incomplete, but the degree of incompleteness is important for long pe-

Table 3. Potentially interesting clusters

Cluster	NgK	V range			V clump
NGC 1193	46	12.	-	17.	
NGC 1245	37	11.	-	14.	13.5
NGC 2141	70	13.	-	16.	14.8
NGC 2158	81	12.	-	16.	15.0
NGC 2204	53	11.	-	14.5	13.8
NGC 2243	25	12.	-	16.	13.8
NGC 2420	45	11.	-	15.	13.0
NGC 2506	45	11.	-	15.	13.0
NGC 2660	39	12.	-	15.	14.2
NGC 6791	80	14.	-	17.5	14.5
NGC 6819	94	12.	-	14.5	13.0
NGC 6939	51	11.	-	13.5	13.0
NGC 7044	83	14.	-	16.5	16.0
Be 32	45	12.	-	16.	13.5
Be 39	57	13.	-	17.	14.2
Mel 66	245	12.	-	16.5	14.5

riod orbits only, as will be shown later on. The examination of the distribution will
start with the (e, log P) diagram. The addition to the discussion of the binaries found
at Victoria in NGC 188 and 7789 would certainly prove interesting.

3.2.1.   The (e, log P) diagram

All 88 red giant stars with a known orbit have been plotted in the (e, log P) diagram
shown in Figure 2. Several general characteristics are well marked:

1) the well-known fact that the shorter periods are circular is readily apparent, but
the limit is not very clear-cut. This is caused by the mixing of all giants, containing
a range in age and mass. Young red giants are more massive and reach larger radii,
they will therefore be circularised at longer periods. Low mass stars, as in M67,
reach also large radii during the ascent of the red giant branch and can also have a
long period and circular orbit. Clump stars of about 2 to 3 $M_\odot$ have comparatively
small radii and the observation shows that clusters with ages around 1 billion years
contain a large proportion of composite binaries and the binaries with the shortest
periods. The existence of orbits with periods less than the cut-off period and non
-zero eccentricities may be explained by the mechanism discussed by Duquennoy et
al. [DMM92] in this volume. It is also worth noting that the star NGC 2360-51

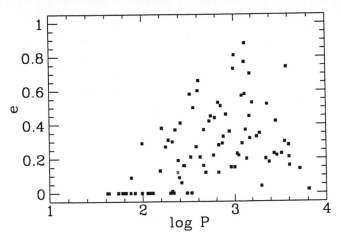

Figure 2. Distribution of the eccentricity as a function of the logarithm of the period. The most remarkable features are the numerous circular orbits and the lack of orbits with $\log P > 2.6$ and $e < 0.10$

which has quite an eccentric orbit for its period ($\log P = 1.995$, $e = 0.29$) occupies an anomalous position in the colour-magnitude diagram of its parent cluster [MeM90].

2) The lack of orbits with periods longer than 250^d ($\log P = 2.4$) and eccentricity smaller than 0.10 is another interesting feature. Due to the number of orbits observed, this cannot result from a bias in the sample. One can therefore conclude that spectroscopic binaries with periods smaller than 3500 to 4000 days are not formed with circular orbits ($e < 0.10$). As shown by Jorissen and Mayor [JoM92] this area is mostly occupied by peculiar giant stars, showing the BaII anomaly. The two objects found in this part of the diagram are IC 4756 # 69 (3.30, 0.04) [MeM90] and M67 # 224 (3.81, 0.00) [MLG90].

3) The shortest period observed for a spectroscopic binary with a red giant primary (excluding stars on the subgiant branch) is 40^d. This value corresponds rather well to the expected critical period from the Roche lobe model [Pla67].

4) The left side of the diagram shows a well defined limit, already seen in the diagram for nearby dwarfs published by Duquennoy and Mayor [DuM91] and among dwarfs in open clusters ([DMM92], this volume). It may result from the evolution of the orbital elements (the change of eccentricity is followed by a change of the period) during the pre-main-sequence phase.

5) The right side of the distribution probably reflects the difficulty of detecting long period binaries with large eccentricity or completing their orbits.

The main feature revealed by this diagram is the lack of short periods commonly found among main sequence binaries. This is easy to show. We have selected, from the entire sample, the binaries with a red giant primary mass in the range of 1.8 to

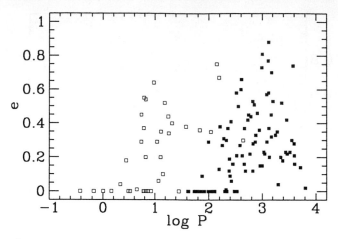

Figure 3. Distribution of the eccentricity as a function of the logarithm of the period for the red giants (filled square) and main sequence stars (open square). The main sequence binaries with log P > 1.2 are nearly completely absent due to lack of observations (P in days)

3.5 $M_\odot$, because the isochrones within the corresponding age interval have a similar morphology. We have looked for the spectroscopic binaries with known orbits among the upper main sequence stars in the same clusters and plot them (open square) together with the red giants (filled square) in a (e, log P) diagram (Figure 3).

Most periods for the B and A main sequence spectroscopic binaries are shorter than log P = 1.4 (25^d) and many of them are circular. There is a wide gap for 1.4 < log P < 2.2, which simply results from the observational conditions and bias in the main sequence sample. Unlike the red giants, none of the main sequence stars have as yet been monitored for several years. Recents works by Morse et al [MML91] and Liu et al [LJB91] have shown the possibility of obtaining accurate radial velocities for B- and A-type stars so that it will be possible to improve this picture and determine the orbits of main sequence spectroscopic binaries in many clusters.

The unsolved question concerns the evolution of the short period binaries and the change of orbital elements when the primary evolves away from the main sequence. What is the proportion of binaries that reaches the red giant stage and what is the fraction that disappears during mass transfer episodes?

3.2.2. Orbital period

Figure 4 shows the distribution of the periods for 88 binaries. The histogram starts decreasing at log P = 3.4, corresponding to P = 2500^d. This value will be used in the following to separate the whole sample into two parts. This is of course a preliminary distribution to illustrate the state of the art. We did not attempt to

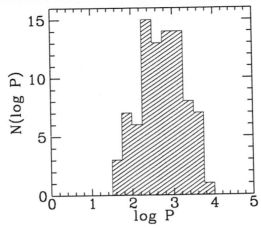

Figure 4. Distribution of the orbital periods for the red giant binaries with known orbits

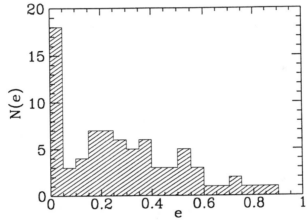

Figure 5. Distribution of the orbital eccentricity for red giant binaries with periods less than 2500^d

model the detection bias and correct the results

3.2.3.   Eccentricity

Because we achieved a different degree of completeness in the orbit determination as a function of the period, we have split the sample into two parts for periods shorter than 2500^d or longer than this value. The first sample should be free of systematic bias, while the latter is certainly biased towards small eccentricity. The distribution of eccentricity for spectroscopic binaries with periods less than 2500^d presented in Figure 5 shows a maximum for $0.20 < e < 0.25$ and a large peak due to the circular orbits ($e = 0$).

3.2.4.   Mass ratio

The addition of the colours of a red giant (primary) and of a main sequence star (secondary) produces remarkable colours for the resulting binary. Such systems are often called composite spectrum binaries, because the observed spectrum is a combination of that from the red giant (although altered) and that of the main sequence star. The representative point in the colour-magnitude diagram often lies close to the middle of the Hertzsprung gap. An example of such a composite binary is shown by Cox 102 [Cox54] in the open cluster NGC 2287. Cox 102 is represented by an open square in Figure 6. The photometric effect produced by the combination of two stars of widely different colours in this plane allows to perform a photometric separation and compute the individual colours of each component.

This has been done with an interactive program belonging to the facilities offered by the database for stars in open clusters described by Mermilliod ([Mer88], [Mer92]). Once the distance modulus has been fixed and an isochrone computed (here for logt = 8.38) from the new grid of Schaller et al [SSM92], the program asks for the position of the binary and an estimate of that of the primary. It then computes the magnitude and colour of the main sequence star and plots, as shown in Figure 6, the position of both components (open circles) joined by a dotted line. The binary loop originates from the red giant magnitude and colour and is computed by adding secondary magnitudes and colours taken from the isochrone (continuous line).

When the correct position of the primary is found (with the help of the isochrone), the binary loop should go through the binary position and the point of the secondary star should be on the main sequence. The solution heavily depends on the choice of the primary colour. If it is too red, then the main sequence component will be too blue. Thus the solution is "unique". Finally, the isochrone also provides an estimate of the mass of both stars. If no photometric effect is visible in the colour magnitude diagram, it is then assumed that the secondary is at least 5 magnitudes fainter. We can therefore derive an upper mass limit for the main sequence companion. Since the spectroscopic orbit gives the minimum mass of the secondary (with the condition that $\sin i = 1$) this often results in a limited mass range for the secondary.

This method has been applied to the spectroscopic binaries in 7 open clusters (NGC 2477, 2539, 2632, 5822, 6633, 6940, and Hyades) which contain several red giants. They all have pretty similar ages, and thus their red giants have closely identical masses. All binary data have been published ([MeM89], [MMA89], [MeM90]), except those for NGC 2477. The result of the analysis is given in Table 4. The columns give successively: the "cluster.star" number, the period, the mass ratio, the red giant mass, the secondary mass when the photometric separation was possible, the maximum and minimum masses in the other cases. The distribution of the mass ratio is presented in Figure 7. In the interval $0.4 < q < 1.0$ the distribution is flat. A similar result has been obtained for the q distribution of the G dwarfs in the Pleiades [MRD92] and the

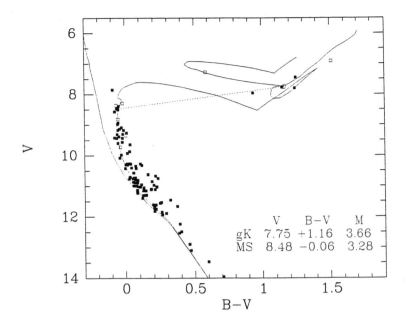

Figure 6. Results of the photometric separation for the composite binary Cox 102 [Cox54] in NGC 2287. The dotted line links the primary and secondary components. Star 102 is located at B-V = 0.58 and V = 7.28

solar vicinity [MGD92]. One can therefore conclude that the mass ratio distribution does not have any maximum for q close to 1.

3.2.5.   Mass function

The distribution of the mass function f(m) for cluster red giants shows a strong maximum for f(m) < 0.10 which is not apparent in similar distributions for dwarf stars. To compare the distribution obtained for red giants and dwarfs in open clusters and in the field, we have defined the relative mass function $f(m)/\mathcal{M}_1$:

$$\frac{f(m)}{\mathcal{M}_1} = \frac{q^3 \sin^3 i}{(1+q)^2} \tag{1}$$

Apart from its dependance on $\sin i$, $f(m)/\mathcal{M}_1$ is solely a function of f(q) and thus independent of the mass M_1 of the primary.

Figure 8 shows the distributions $f(m)/\mathcal{M}_1$ vs $\log P$ for three different (unbiased) samples with selected primary mass M_1. The red giant sample $(\overline{\mathcal{M}_1} = 2.7\mathcal{M}_\odot)$ presents an excess of companions (compared to dwarfs with $\overline{\mathcal{M}_1} = 1\mathcal{M}_\odot$) with $f(m)/\mathcal{M}_1 < 0.015$ or $M_2 < 0.8\mathcal{M}_\odot$. Such an excess of companions with mass

Table 4. Individual masses of the binary components

Cluster	P	q	M_1	M_2	$M_{2,max}$	$M_{2,min}$
2447.0025	113.80	0.63	2.89	1.82		1.58
2477.1025	41.56		2.06		0.86	0.34
2477.1044	2955.29		2.19		0.95	0.44
2477.2204	1317.42	0.72	2.06	1.48		
2477.3003	1775.49	0.60	2.06	1.23		
2477.3176	276.20		2.05		0.83	0.43
2477.5073	326.09		2.08		0.89	0.79
2477.6020	226.16		2.06		0.86	
2477.6062	479.73	0.73	2.06	1.51		
2477.8017	60.32		2.05		0.83	0.81
2539.0032	242.05	0.56	2.42	1.36	0.99	
2539.0038	585.2	0.86	2.28	1.96		
2539.0042	404.97		2.31		0.90	0.45
2632.0428	998.3		2.26		0.90	0.52
5822.0002	1001.2	0.72	2.34	1.69		
5822.0011	877.9		2.17		0.83	0.66
5822.0080	65.45		2.20		0.92	0.92
5822.0151	1374.7		2.17		0.85	0.39
5822.0312	978.18	0.82	2.24	1.84		
6633.0070	1224.3	0.79	2.43	1.92		
6940.0084	57.17	0.85	2.14	1.81		
6940.0092	549.2		2.14		0.82	0.31
6940.0100	82.53	0.88	2.16	1.90		
6940.0111	3593.		2.14		0.86	0.55
6940.0130	281.71		2.14		0.88	0.60
6940.0189	210.73	0.91	2.21	2.01		
5025.0041	532.		2.29		0.92	0.20

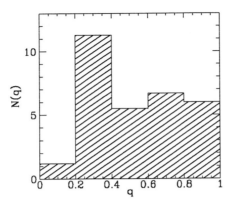

Figure 7. Distribution of the mass ratio for primaries with $2 < M_1 < 3M_\odot$. The number obtained for q < 0.2 is not significant

$M_2 < 0.8M_\odot$ could result either from the presence of white dwarf or low-mass main sequence secondaries. If the secondaries are predominently degenerate companions resulting from the past evolution of the more massive original primary, we should observe a difference between the $(e, \log P)$ diagrams for stars with $f(m)/M_1 < 0.01$ or with $f(m)/M_1 > 0.02$. Such a difference may result from wind accretion (see Boffin and Jorissen [BoJ88], and the study of S stars by Jorissen and Mayor [JoM92]). No difference is observed when we compare the $(e, \log P)$ diagrams for samples with different values of the ratio $f(m)/M_1$. (Figure 9). We then conclude that the excess of companions with $M_2 < 0.8M_\odot$ is most probably due to low-mass main sequence secondaries. This however does not exclude the existence of white dwarfs in binaries with a red giant primary. The Kolmogorov - Smirnov test applied to the cumulative distributions of $f(m)/M_1$ for the cluster red giants and dwarfs shows that the two distributions are different at the 3% level.

4.   Conclusions

The subsample of 638 red giant members extracted from the total observing list of 1048 stars in 177 open clusters gives a spectroscopic binary ($P < 10^4 d$) frequency equal to 23% (144/638). This frequency is stable against the definition of the sample, and is not influenced by the inclusion or not of NGC 2477. The variation of the number of binaries from cluster to cluster is normally comprised within the 1 σ limit. However, there are a few exceptions to this rule (NGC 188 and 6705), which shows that the binary frequency may vary from cluster to cluster. Observations of more old and rich open clusters would be very useful to investigate this particular point.

Several examples show that the census of red giant stars in several open clusters is probably not complete. New members have been found in the unstudied outer parts

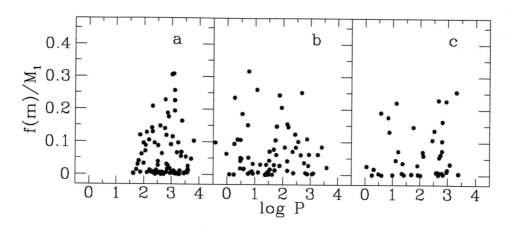

Figure 8. Distribution of the relative mass function for: (a) red giants; (b) cluster F-G dwarfs, (c) field dwarfs

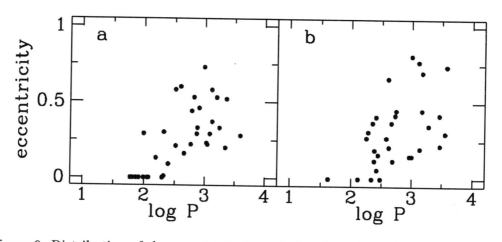

Figure 9. Distribution of the eccentricity for red giant binaries with (a) $f(m)/\mathcal{M}_1 > 0.02$; (b) $f(m)/\mathcal{M}_1 < 0.01$

(i.e. outside 1 cluster radius) of a couple of open clusters. The detection of all red giants in open cluster fields is important to define an unbiased and complete sample. This condition is especially important for a study like this one.

Since the present analysis is based on a sample of determined orbits which represents about half of the total number of binaries discovered, the distributions discussed are preliminary ones. However, the distributions for stars with periods less than 2500^d will probably not change much. The fact that there are no orbits with periods longer than 250^d and eccentricity less than 0.10 does not depend on the sample completeness. We can already claim that binaries with periods longer than 250^d are not formed in circular orbits.

The distribution of mass ratios in the interval of primary mass $2 < M_1 < 3 \, \mathcal{M}_\odot$ is flat for $0.4 < q < 1.0$. Furthermore, the distribution of the mass function f(m) may depend on the mass of the primary.

5.   Acknowledgements

We are grateful to Dr R. McClure (Victoria) for giving us information on the number of binaries found in NGC 188 and 7789. The long term Coravel observing program has been realised thanks to continuous grants of the Swiss National Funds for Scientific Research (FNRS).

References

[APT89] Anthony-Twarog B.J., Payne D.M, Twarog B.A. *A BV photographic study of the southern cluster, NGC 6259. AJ 97, 1048-1063*

[BoJ88] Boffin H., Jorissen, A. *Can a Barium star be produced by wind accretion in a detached binary? AA 205, 155-163*

[CLB91] Claria J.J., Lapasset E., Bosio M.A. *Membership, fundamental parameters, and luminosity function of the open cluster NGC 5662. 1991, MNRAS 249, 193-207*

[ClM92] Claria J.J., Mermilliod J.-C. *Membership, binarity and metallicity of red giants in the open cluster NGC 6134. 1992, AAS (in press)*

[Cox54] Cox A.N. *A study of the galactic cluster NGC 2287. ApJ 119, 188-196*

[DaK89] Dachs J., Kabus H. *UBV photometry and the structure of the galactic cluster NGC 2516. AAS 78, 25-49*

[DuM91] Duquennoy A., Mayor M. *Multiplicity among solar-type stars in the solar neighbourhood. II. Distribution of the orbital elements in an unbiased sample. 1991, AA 248, 485*

[DMM92] Duquennoy A., Mayor M., Mermilliod J.-C. *Distribution and evolution of orbital elements for 1 $M_\odot$ Primaries. 1992, in Binaries as Tracers of Stellar Formation, Eds A. Duquennoy and M. Mayor, Cambridge, Univ. Press, in press*

[GrG77] Griffin R.F., Gunn J.E. *Hyades giants δ and θ^1 Tauri as spectroscopic binaries. 1977, AJ 82, 176-178*

[JoM92] Jorissen A., Mayor M. *Orbital elements of S stars: revisiting the evolutionary status of S stars. 1992, AA (in press) (see also ESO preprint 817)*

[HaM83] Harris H.C., McClure R.D. *Radial velocities of a random sample of K giant stars and implications concerning multiplicity among giant stars in clusters. ApJ 265, L77-81*

[HaM85] Harris H.C., McClure R.D. *Dynamics of NGC 188. in Stellar Radial velocities, Eds A.G.D. Philip and D.W. Latham (Schenectady, Davis Press) p. 257-261*

[LML89] Lee C.W., Mathieu R.D., Latham D.W. *Two spectroscopic binaries in the open cluster M11 and their implication for the cluster distance modulus. 1989, AJ 97, 1710-1715*

[LJB91] Liu T., Janes K.A., Bania T.M. *Radial velocity measurements in the Pleiades. 1991, ApJ 377, 141-149*

[Mat84] Mathieu R.D. *The structure and dynamics of the open cluster M11. 1984, ApJ 284, 643-662*

[MLG86] Mathieu R.D., Latham D.W., Griffin R.F.,Gunn J.E. *Precise radial velocities of late-type stars in the open clusters M11 and M67. 1986, AJ 92, 1100-117*

[MLG90] Mathieu R.D., Latham D.W., Griffin R.F. *Orbits of 22 spectroscopic binaries in the open cluster M67. 1990, AJ 100, 1859-1881*

[McC92] McClure R.D. *1992, private communication*

[MGD92] Mazeh, T., Goldberg, D., Duquennoy, A., Mayor, M. *On the mass-ratio distribution of spectroscopic binaries with solar-type primaries. 1992, ApJ (in press)*

[Mer88] Mermilliod J.-C. *Description of a database for stars in open clusters. 1988, Inform. Bull. CDS 35, 77-91*

[Mer92] Mermilliod J.-C. *The database for stars in open clusters. II. A progress report on the introduction of new data. 1992, Inform. Bull. CDS 40, 115-120*

[MeM89] Mermilliod J.-C., Mayor M. *Membership and binarity of red giants in open clusters determined by photoelectric radial velocities. 1980, in Star Clusters, IAU Sym. no 85, Ed. J.E. Hesser (Reidel) p. 361-362*

[MeM89] Mermilliod J.-C., Mayor M. *Red giants in open clusters. I. Binarity and stellar evolution in five Hyades generation clusters: NGC 2447, 2539, 2632, 6633 and 6940.* 1989, AA 219, 125-141

[MeM90] Mermilliod J.-C., Mayor M. *Red giants in open clusters. III. Binarity and stellar evolution in five intermediate-age clusters: NGC 2360, 2423, 5822, 6811 and IC 4756.* 1990, AA 237, 61-72

[MMA89] Mermilliod J.-C., Mayor M., Andersen J., Norstrom B., Lindgren H., Duquennoy A. *Red giants in open clusters. II. Orbits for ten spectroscopic binaries in NGC 2360, 2437, 2447, 5822, 5823, and 6475.* 1989, AAS 79, 11-18

[MMB87] Mermilliod J.-C., Mayor M., Burki G. *Membership of cepheids and red giants in 8 open clusters: NGC 129, 6067,6087, 6649, IC 4725, Ly 6, Ru 79.* 1987, AAS 70, 389

[MRD92] Mermilliod J.-C., Rosvick J., Duquennoy A., Mayor M. *Investigation of the Pleiades cluster. II. Binary stars in the F5-K0 spectral region.* 1992, AA (submitted)

[MML91] Morse A.J., Mathieu R.D., Levine S.E. *The measurement of precise radial velocities of early-type stars.* 1991, AJ 101, 1495-1510

[Pla67] Plavec M. *On the problem of contact primaries in semi-detached close binary systems.* 1967, BAC 18, 253

[SSM92] Schaller G., Schaerer D., Meynet G., Maeder A. *New grids of stellar models from 0.8 to 120 $M_\odot$ at Z=0.020 and Z=0.001.* 1992, AA (submitted)

[Vin40] VinterHansen J.M. *The spectroscopic binary 12 Comae Berenices.* 1940, Lick Obs. Bull. 19, 101-103

Are Barium Dwarfs Progenitors of Barium Giants ?

Pierre North *   *Antoine Duquennoy* [†]

Abstract

The results of a radial velocity monitoring of a sample of nine Ba dwarfs are presented. The high rate of binaries among this sample, together with the periods, eccentricities and mass functions found for the few objects with enough phase coverage, is quite compatible with the idea that these Ba dwarfs may be progenitors of the Ba giants.

On the other hand, the bright, mild Ba dwarf HR 107 recently discovered had a constant radial velocity for years.

1.   Introduction

Most Ba II and CH stars are yellow giants with an overabundance of carbon and of s-process elements such as Sr, Y, Zr, Ba, La etc. Theoretical calculations (see [IR83]) show that these elements can indeed be produced during the very late phases of stellar evolution and are brought to the surface by the third dredge-up. However, the production can only begin in the AGB phase, during the thermal pulses, while the Ba II giants have luminosities typical of the "clump" and are therefore not luminous enough to generate s-process elements themselves.

This paradox was solved by the discovery that all barium giants are SB1 binaries (see [MC80], [MC83], [JM88]), and that their companions are likely to be white dwarfs (according to the mass functions obtained by McClure and Woodsworth [MC90]) or are indeed so (UV detection by Böhm-Vitense et al. [BN84]). The white dwarf is supposed to be the remnant of the former primary of the binary system, which went through the AGB phase and lost its envelope, thereby polluting its companion (the present primary) with material enriched in carbon and s-process elements.

The interesting point here, is that the former secondary may have been polluted while still being on the main sequence, because the accretion cross-section does not

---

*Institut d'Astronomie de l'Université de Lausanne, CH–1290 Chavannes-des-Bois, Switzerland.

[†]Observatoire de Geneve, CH–1290 Sauverny, Switzerland.

depend on the radius of the star, as shown by Boffin and Jorissen [BJ88]. Therefore, the mass-transfer scenario just described predicts the existence, and maybe even the predominence, of barium *dwarfs*. The latter, however, have only been recognized as such recently, in spite of the early discovery by Bond [B74] of the "CH subgiants", several of which being dwarfs rather than subgiants according to later, detailed studies by Luck and Bond [LB82], [LB91]. An important reason which prevented the CH subgiants from being considered the main sequence counterparts of the Ba II giants (sharing with them the same mass-transfer history) is that their spectral types were all confined within a narrow interval ($\pm$ 2 or so subtypes) centered on G0. This was impossible to reconcile with the more or less random distribution of types (all along the main sequence) expected from the mass-transfer hypothesis. However, hotter Ba dwarfs, with spectral types around F5, as well as cooler ones, have been discovered recently (see [TL89], [ND91a], [ND91b], [L91], [LB91], [NB92]), which increases the relevance and the value of the mass- transfer scenario not only for the dwarfs, but also for the giants.

Because of the rather late recognition of the existence of barium dwarfs, no radial velocity monitoring of them has yet been completed, although some preliminary results about CH subgiants have been published by McClure [MC85]. The radial velocity results presented here concern a small sample of nine stars selected from what appeared to be a new kind of chemically peculiar stars discovered by Bidelman on objective-prism plates, the "F str λ4077" stars (see [B81], [B83], [B85]). One of these stars, HD 182274, was already known as a "CH subgiant" (see [LB82]) and has a variable radial velocity according to McClure [MC85], while the other eight had their Ba peculiarity confirmed by high resolution spectroscopy (see [NB92]). They are all dwarfs according to Geneva photometry.

The results are presented in the fourth section for the sample of nine stars discovered by Bidelman, and also for the bright Ba dwarf HR 107 discovered by Tomkin et al. [TL89].

2.   Observations

The radial velocity observations have been (and are still being) made with the 1 m Swiss telescope equipped with the Coravel scanner. They began in 1987. The precision of the observations is between 0.4 and 0.7 km s^{-1}, depending on the effective temperature and on the projected rotational velocity of the star.

The star HR 107 had been observed a few times in 1982 and 1983 (the barium anomaly of this star was then unknown), then from 1989 on, after a six-years gap.

The photometric observations have been made both from La Silla, Chile, with the 0.7 m Swiss telescope and from Jungfraujoch, Switzerland, with the 0.76 m telescope. They were made in the Geneva system and have been interpreted and published by

North and Duquennoy [ND91a].

3.   The evolutionary state of the observed stars

It is important here to settle the evolutionary state of the stars we are speaking about, and to show they indeed belong to the main sequence, not to the giant branch. The photometric data, combined with a calibration in terms of effective temperature and surface gravity, yield surface gravities greater than log g = 4.0 for all but one of our nine stars, the only exception being HD 107574 with log g = 3.87 (see [ND91a]).

Another evidence is given here on the basis of more accurate photometric colours, and using the concept of "photometric boxes" defined by Golay et al. [GP69], [GM77]. According to this concept, stars which share almost the same photometric colours have almost identical physical properties. Using a programme kindly made available by Bernard Nicolet (Geneva Observatory), we took the census of stars sharing the same Geneva colour indices as our barium stars, within 0.012 mag. Keeping only those with a good MK classification (from the work of Nancy Houk [HC75], [H78], [H82], [H88] in most cases), we see from Table 1 that *all* nine barium dwarfs are photometrically identical with luminosity class V stars. The same is true of the bright, mild Ba dwarf HR 107, which anyway was already classified F5V by Anne Cowley [C76]. Therefore, all Ba stars considered here clearly belong to the main sequence.

Table 1. List of stars falling in the same photometric "box" as the barium stars (in boldface), with a box "radius" of 0.$^{\rm m}$012. The qualities (third column) are those of Houk when mentioned; a "B" in this column indicates that the MK type has been taken from Dorit Hoffleit's Bright Star Catalogue.

HD/BD	MK	Q	V	[U-B]	[V-B]	[B1-B]	[B2-B]	[V1-B]	[G-B]
15306			8.934	1.339	0.492	0.983	1.397	1.216	1.583
29992	F3V	3	5.024	1.350	0.492	0.982	1.387	1.213	1.579
48565			7.212	1.181	0.298	1.007	1.355	1.028	1.352
126793	G0wF3/5	1	8.177	1.180	0.297	1.008	1.361	1.025	1.352
3119	F6/7V	2	9.884	1.179	0.288	1.016	1.357	1.028	1.345
92545			8.545	1.278	0.344	1.016	1.362	1.080	1.412
5912	F5V	1	9.507	1.277	0.343	1.011	1.366	1.077	1.408
8471	F6V	2	9.446	1.279	0.351	1.019	1.367	1.083	1.419
23308	F7V	1	6.491	1.269	0.346	1.022	1.355	1.078	1.411
146499	G0V	2	9.706	1.272	0.341	1.024	1.353	1.074	1.403
183577	F6V	2	6.469	1.271	0.338	1.018	1.361	1.071	1.403

Table 1. (continued)

HD/BD	MK	Q	V	[U-B]	[V-B]	[B1-B]	[B2-B]	[V1-B]	[G-B]
92545	(cont.)		8.545	1.278	0.344	1.016	1.362	1.080	1.412
16878	F5V	1	8.030	1.287	0.340	1.024	1.356	1.082	1.410
5204	F8V	1	9.685	1.287	0.347	1.024	1.353	1.077	1.405
196235	F5V	1	8.123	1.268	0.350	1.013	1.371	1.079	1.415
120691	F7/8V	1	7.170	1.288	0.342	1.022	1.361	1.078	1.407
6402	F6V	1	8.242	1.289	0.344	1.022	1.358	1.079	1.409
134483	F6V	1	6.668	1.278	0.344	1.022	1.353	1.084	1.423
4530	F7V	1	9.056	1.277	0.338	1.027	1.358	1.074	1.405
204554	F7V	1	8.363	1.276	0.341	1.024	1.350	1.074	1.403
16609	F7V	2	7.963	1.266	0.336	1.016	1.357	1.073	1.406
106191			10.025	1.255	0.231	1.053	1.330	0.966	1.284
143846	G2V	2	7.844	1.248	0.227	1.047	1.325	0.971	1.279
214216	G2/3V	1	9.512	1.247	0.230	1.050	1.337	0.970	1.276
27998	G3V	2	9.230	1.255	0.230	1.052	1.333	0.968	1.275
16519	G0V	2	9.236	1.250	0.234	1.056	1.331	0.975	1.283
211332	G3V	1	8.477	1.257	0.233	1.045	1.330	0.976	1.278
30361	G0V	1	8.339	1.265	0.226	1.045	1.331	0.967	1.275
73744	G2V	1	7.591	1.244	0.227	1.055	1.331	0.972	1.277
107574			8.545	1.291	0.426	0.983	1.384	1.153	1.505
86847	F2V	1	8.990	1.287	0.416	0.979	1.391	1.153	1.493
147609			9.188	1.247	0.348	1.002	1.367	1.094	1.435
400	F6IV	B	6.225	1.248	0.352	1.010	1.364	1.093	1.436
118261	F6V	B	5.621	1.254	0.357	1.013	1.362	1.096	1.437
182274			7.797	1.204	0.368	1.005	1.369	1.099	1.431
24115	F6wF3V	1	8.937	1.213	0.370	0.998	1.374	1.096	1.428
33476	F6/7V	1	7.708	1.214	0.370	1.002	1.373	1.104	1.440
221531			8.324	1.263	0.423	0.994	1.380	1.152	1.505
201772	F5V	1	5.239	1.261	0.428	0.994	1.380	1.156	1.508
12586	F5/6V	1	7.687	1.258	0.426	1.000	1.377	1.154	1.507
96143	F5/6V	2	7.611	1.261	0.428	0.998	1.382	1.158	1.506
739	F3/5V	1	5.230	1.257	0.427	0.997	1.378	1.157	1.509
183028	F5V	1	6.279	1.266	0.426	0.996	1.373	1.154	1.508
4910	F5V	2	8.449	1.268	0.431	0.996	1.377	1.160	1.512
6368	F5V	1	8.773	1.256	0.432	0.991	1.376	1.156	1.511
162521	F5V	1	6.352	1.254	0.421	0.997	1.373	1.148	1.499
799	F3V	2	9.878	1.256	0.434	0.990	1.386	1.163	1.508
28246	F5IV/V	1	6.367	1.269	0.422	1.005	1.374	1.152	1.505
165069	F5V	1	8.008	1.252	0.423	0.998	1.384	1.145	1.504
31204	F4V:	B	6.402	1.251	0.421	0.999	1.382	1.152	1.495

Table 1. (continued)

HD/BD	MK	Q	V	[U-B]	[V-B]	[B1-B]	[B2-B]	[V1-B]	[G-B]
+18°5215			9.695	1.185	0.377	0.991	1.368	1.110	1.452
26153	F5V	1	8.649	1.180	0.385	0.989	1.377	1.116	1.456
HR 107	F5V Ba		6.045	1.263	0.444	0.992	1.401	1.172	1.525
129698	F5V		6.627	1.257	0.442	0.991	1.389	1.171	1.525
33256	F5V		5.110	1.271	0.432	0.995	1.408	1.174	1.529

4. Radial velocities and orbits

The results of the radial velocity measurements are displayed in Figures 1 to 9, either as a function of time when no period could be obtained, or as a function of phase. In the latter case, the fitted curve is superimposed. Because of the great length of the periods, we could find an orbit in only three cases. The corresponding orbital parameters are listed in Table 2. Only one of Bidelman's stars, HD 92545, does not vary in a very significant way, although the last measurement (not yet very precise: only a preliminary reduction has been done in dome) suggests a trend. Until very recently, this seemed to be the case of HD 182274 too, although this star was found variable by McClure (see Fig. 3 of [MC85]). Thanks to two early measurements made in the context of another programme, and to the few most recent ones, the variability of HD 182274 is confirmed at last. However, McClure's measurement at about JD 2444980 and at $V_r \approx -8$ kms^{-1} cannot be so easily reconciled with our data.

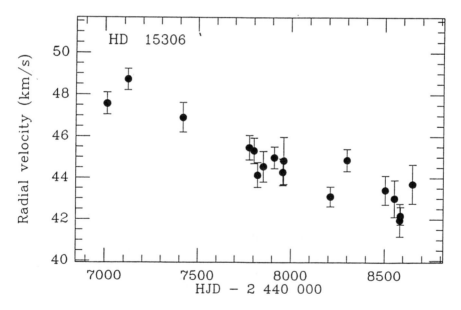

Fig. 1: Radial velocity variation of HD 15306.

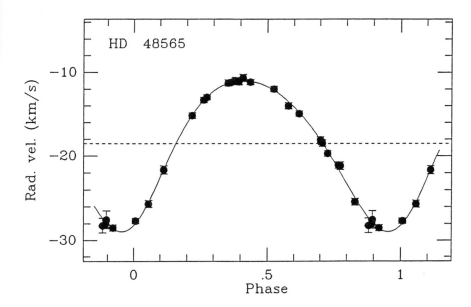

Fig. 2: Radial velocity curve of HD 48565.

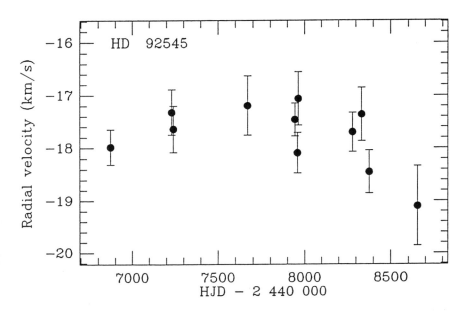

Fig. 3: Radial velocity results for HD 92545.

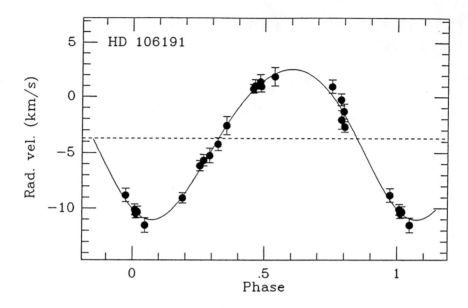

Fig. 4: Radial velocity curve of HD 106191.

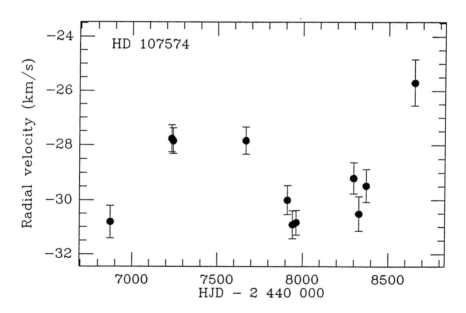

Fig. 5: Radial velocity variation of HD 107574

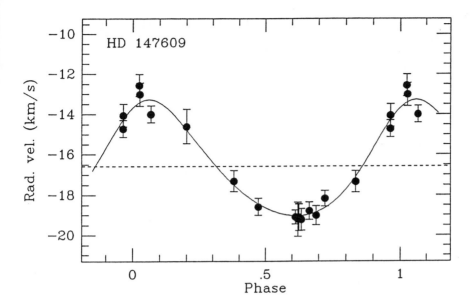

Fig. 6: Radial velocity curve of HD 147609.

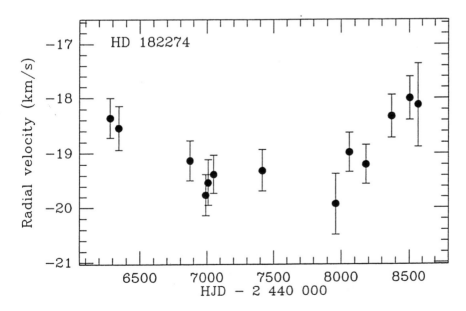

Fig. 7: Radial velocity variation of HD 182274.

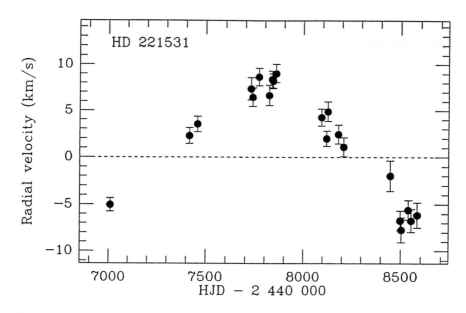

Fig. 8: Radial velocity variation of HD 221531.

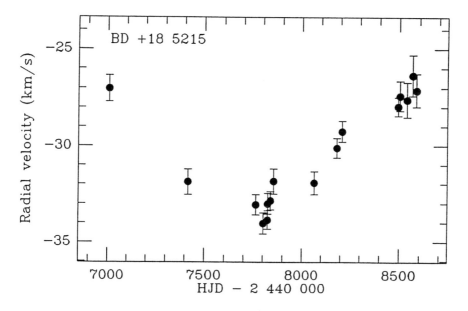

Fig. 9: Radial velocity variation of BD +18°5215.

Table 2. Orbital parameters of barium dwarfs with enough phase coverage.

Parameter	HD 48565		HD 106191		HD 147609	
P (days)	73.28	±0.03	1316	±20	1208	±57
$T_o - 2440000$	7748.3	±0.7	7946	±128	7205	±132
e	0.177	±0.014	0.09	±0.04	0.17	±0.13
V_o (kms^{-1})	-18.51	±0.07	-3.72	±0.14	-16.59	±0.20
ω_1 (°)	203.7	±3.4	149	±35	329	±32
K_1 (kms^{-1})	9.00	±0.12	6.80	±0.20	2.88	±0.17
$a_1 \sin i$ (10^6 km)	8.92	±0.14	123	±6	47	±6
f(m)	0.00528	±0.00025	0.043	±0.005	0.0029	±0.0008
n	26		21		16	
σ_{res}	0.32		0.57		0.46	

The rate of binaries among the nine barium stars taken from Bidelman's sample is
eight out of nine, i.e. 89%. It is possible that the remaining apparently stable object
proves variable after a season more of monitoring. Such a high rate of binaries is
quite consistent with the idea that these barium dwarfs are indeed related with the
Ba giants, and may even be their progenitors. Furthermore, all orbital periods are
longer than 73 days, and are typically in the order of 1000 or 2000 days, while the
periods of the Ba giants range from 81 days to 20 years or so. This is again very
consistent.

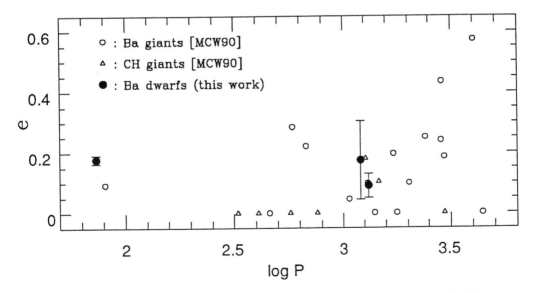

Fig. 10: e vs log P for Ba dwarfs (full dots, this work) and for Ba and CH giants
([MC90], open dots and triangles respectively).

The orbital parameters are summarized and compared with those of the Ba giants in the e vs log P diagram of Figure 10. In this diagram we show the recent results of McClure and Woodsworth [MC90] for the Ba giants – indicated by open dots – and for the CH giants (which seem to be, roughly speaking, the Pop. II counterpart of the Ba stars) which are indicated by open triangles. The number of dwarfs with determined orbits is uncomfortably small and we admit that Figure 10 does not have a very compelling statistical value. Nevertheless, it is rather certain from Figures 1, 5, 7, 8 and 9 that as soon as enough data will have been gathered, the five remaining points will have log P > 3.0 and e < 0.4 or so and will probably lie well within the domain of the Ba giants in Figure 10. The mass functions are also quite similar with those of the system containing a barium or a CH giant [MC90].

A comparison with the *normal,* main sequence G-type binaries is provided in Figure 11: our three barium dwarfs (full dots with error bars) have been added to the e-log P diagram already published by Duquennoy and Mayor (see Figure 5 of [DM91]; note that Fig. 11 is slightly different from Fig. 5 of [DM91], because the latter was accidentally built from a file containing obsolete data for 6 binaries; the correct data are to be taken from Tables 2 and 3 of [DM91]). In spite of the small number of Ba dwarfs, the difference between their eccentricities and those of the normal G dwarfs is quite striking. Clearly, the orbits of the former must have been modified by some dissipative process since the formation of these binary systems, and a mass-transfer event would account naturally for this.

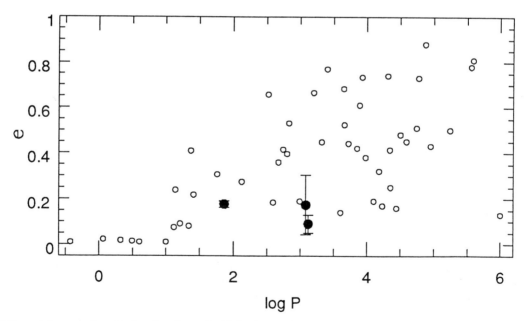

Fig. 11: e vs log P for Ba dwarfs (full dots, this work) and for the nearby G-dwarf sample of Duquennoy and Mayor (open dots, [DM91])

Together with the overabundance of the s-process elements [NB92], Figure 11 provides a strong confirmation of the mass-transfer scenario for the barium dwarfs.

The star HR 107 discovered by Tomkin et al. [TL89] had been measured a long time ago with Coravel. Unfortunately, there is a six years gap in the data, since the measurements were only resumed in 1989, after the discovery of the barium peculiarity. The Coravel data are displayed in Figure 12.

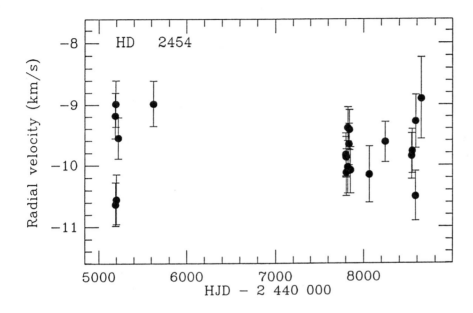

Fig. 12: Radial velocity results for HR 107 = HD 2454.

The scatter of the V_r values seemed a little larger than the internal error, especially for the early measurements, which suggested short timescale variation. This is why we have measured this star 16 times over three and a half hours in the night of November 24, 1991 (represented by one mean point on JD 8585). The result is clearly negative, since the r.m.s. scatter is only 0.33 kms^{-1}, which fits very well the expected internal standard deviation, so there is no variation (due e.g. to pulsation) at least on this timescale. The high scatter of the early measurements is therefore probably not meaningful.

We include this star into our paper because of its brightness and because it was studied in detail [TL89]. It is rather disappointing to see how constant its radial velocity has remained during more than nine years: $< V_r > = -9.7$ kms^{-1}, with a standard deviation of ± 0.51 kms^{-1}. It is the brightest barium dwarf published to date (apart from the strange cases of ϵ Ind, type K5V found by Kollatschny [K80] and ξ Boo A, type G8V found by Boyarchuk and Eglitis [BE84], where the pattern of abundances is very different from that of a typical giant Ba II star, as shown by Lambert [L91]), and no sign of variability can be found either in published data. Anderson and Kraft [AK72] made seven measurements over two years and obtained a mean value $< V_r >= -8.5 \ \text{kms}^{-1}$, with a standard deviation of $\pm 0.31 \ \text{kms}^{-1}$. Although this result is $1.2 \ \text{kms}^{-1}$ higher than the mean of our Coravel observations, we do not think it is due to any long-term variability, because there may well be some systematic shift between the radial velocities obtained at the Lick 120-inch reflector and ours. In any case, it looks like the star HR 107 has remained stable within $\pm 1 \ \text{kms}^{-1}$ for the last 22 years at least! Of course, one may imagine that HR 107 is a binary with a long period and a very excentric orbit, where significant variations of the radial velocity occurs on a small phase range only and have therefore not yet been detected. Another possibility is that the orbit is seen nearly pole-on. We continue the observations in order to obtain a homogeneous set of data spanning a larger time interval.

5.   Conclusion

We have shown that not only the rate of binaries, but also the orbital characteristics of our small sample of nine barium dwarfs are completely similar with those concerning the barium and CH giants. Therefore, from the dynamical point of view alone, it seems that the answer to our title must be positive. In other words, if the mass-transfer scenario is valid, then one can admit that the mass-transfer very often occurs when the gaining star is still on the main sequence, as foreseen by Boffin and Jorissen [BJ88]. When the transfer is over and the evolved star has become a white dwarf, the former secondary has become the primary component of the binary system and is seen as a barium dwarf. Later, this star evolves and becomes a Ba (or CH, depending on the metallicity) giant. This does not exclude the possibility, of course, that some barium giants have undergone the mass-transfer while they were *already* on the red giant branch and are, in this sense, "primordial" Ba giants. But our results tend to show that at least part of the Ba giants are not "primordial" ones, but had been Ba dwarfs before.

The question is still left open, whether our dwarfs are progenitors of *mild* or of *classical* Ba giants. This question can only be answered by detailed abundance analyses, which are underway, and also by future progresses in stellar evolution theory and in modelling the mass-transfer event.

Acknowledgements: We thank the observers who kindly contributed to the monitoring of our stars, especially Dr. Jean-Claude Mermilliod, Mr. Bernard Pernier and Dr. Jose Renan de Medeiros. We thank also Dr. Bernard Nicolet for allowing us to use his programme of photometric boxes, and Drs. Alain Jorissen and Michel Mayor for stimulating discussions. This work was supported in part by the Swiss National Foundation for Scientific Research.

References

[AK72]   Anderson K.S., Kraft R.P., 1972, ApJ 172, 631

[B81]   Bidelman W.P., 1981, AJ 86, 553

[B83]   Bidelman W.P., 1983, AJ 88, 1182

[B85]   Bidelman W.P., 1985, AJ 90, 341

[BJ88]   Boffin H.M.J., Jorissen A., 1988, A&A 205, 155

[BN84]   Böhm-Vitense, E., Nemec, J., Profitt, C., 1984, ApJ 278, 726

[B74]   Bond H.E., 1974, ApJ 194, 95

[BE84]   Boyarchuk A.A., Eglitis I., 1984, Bull. Crimean Astrophys. Obs. 64, 11

[C76]   Cowley A.P., 1976, PASP 88, 95

[DM91]   Duquennoy A., Mayor M., 1991, A&A 248, 485

[GP69]   Golay M., Peytremann E., Maeder A., 1969, Publ. Obs. Genève, Série A, No 76, 44

[GM77]   Golay M., Mandwewala N., Bartholdi P., 1977, A&A 60, 181

[HC75]   Houk N., Cowley A.P., 1975, *University of Michigan Catalogue of two-dimensional spectral types for the HD stars, Vol. 1. University of Michigan, Ann Arbor*

[H78]   Houk N., 1978, *ibidem, Vol. 2.*

[H82]   Houk N., 1982, *ibidem, Vol. 3.*

[H88]   Houk N., Smith-Moore M., 1988, *ibidem, Vol. 4*

[IR83]   Iben I. Jr., Renzini A., 1983, ARA&A 21, 271

[JM88]   Jorissen A., Mayor M., 1988, A&A 198, 187

[K80]   Kollatschny W., 1980, A&A 86, 308

[L85]   Lambert D.L., 1985, in: Jaschek M., Keenan P.C. (eds.) *Cool Stars with Excesses of Heavy Elements. Reidel, Dordrecht, p. 191.*

[LB82]   Luck R.E., Bond H.E., 1982, ApJ 259, 792

[LB91]   Luck R.E., Bond H.E., 1991, ApJS 77, 515

[L91]   Lü P.K., 1991, AJ 101, 2229

[MC83]   McClure R.D., 1983, ApJ 268, 264

[MC85]   McClure R.D., 1985, in: Jaschek M., Keenan P.C. (eds.) *Cool Stars with Excesses of Heavy Elements*. Reidel, Dordrecht, p. 315.

[MC90]   McClure R.D., Woodsworth A.W., 1990, ApJ 352, 709

[MC80]   McClure R.D., Fletcher J.M., Nemec J.M., 1980, ApJ 238, L35

[N89]   Nissen P.E., 1989, A&A 219, L15

[ND91a] North P., Duquennoy A., 1991a, A&A 244, 335

[ND91b] North P., Duquennoy A., 1991b, in: Michaud G., Tutukov A., Bergevin M. (eds.) *Evolution of Stars: the Photospheric Abundance Connection. Poster papers presented at the 145th Symposium of the IAU. Michaud G., Montréal*, p. 39.

[NB92]   North P., Berthet S., Lanz T., 1992, A&A (in preparation)

[TL89]   Tomkin J., Lambert D.L., Edvardsson B., Gustaffson B., Nissen P.E., 1989, A&A 219, L15

RZ Eridani as a constraint on synchronization and circularization times

Pierre North *   *Gilbert Burki* [†]   *Zdenek Kvíz* [‡]

Abstract

The RS CVn - type eclipsing binary RZ Eri has been monitored with Geneva photometry for twelve years. These photometric data allowed a precise estimate of the parameters of the system, which has an eccentric orbit, and to show that the cooler component is not pseudo-synchronized. This system is compared with three other similar ones and a preliminary discussion of its tidal evolution is proposed.

1.   Introduction

Different theories of tidal evolution have been proposed in the recent past, especially by Zahn (see [Z77] and [Z89]), by Tassoul (see [T87] and [T88]) and by Hut (see [H80], [H81]). The tidal effects have essentially two consequences which can be readily tested by observation:

– Each component tends to have its axial rotation synchronized with its orbital rotation, like the Moon in the Earth - Moon system.

– The eccentricity of the orbit tends towards zero.

Synchronization and circularization are achieved after a characteristic timescale which is generally much longer in the latter case than in the former, and which can be foreseen theoretically.

There are several possible observational tests of these theories: the circularization time may be estimated by observing many binaries with different periods and eccentricities, and by determining the critical period under which all binaries have a circular orbit (see e.g. [MM84]). This is ideally valid when all considered binaries

---

*Institut d'Astronomie de l'Université de Lausanne, CH–1290 Chavannes-des-Bois, Switzerland.

[†]Observatoire de Genève, CH–1290 Sauverny, Switzerland.

[‡]School of Physics, University of New South Wales, Kensington NSW 2033, Australia, and Observatoire de Genève.

have the same age. Another philosophy consists in using eclipsing systems which are also double-line spectroscopic binaries, for which every fundamental parameter can therefore be known. One can then try to reproduce the tidal history of the system up to the present time, starting from the theory, and compare the result with the observations. Examples of this approach can be found in works by Habets and Zwaan [HZ89] and Hall and Henry [HH90].

In this work, we want to draw the attention to an interesting system, the characteristics of which have recently been determined with a much improved precision thanks to a long-term photometric monitoring. This system is RZ Eridani and contains an A8-F0IV secondary which is just leaving the main sequence, and a G8-K0IV-III primary which is ascending the red giant branch. It is eclipsing, which allows a precise determination of the radii of the components and of the orbital eccentricity, and is an SB2 system as well, so that the individual masses can be determined. Furthermore, it is an RS CVn system according to the definition of Fekel et al. [FMH86], i.e. the cooler star is chromospherically active, which allows a precise determination of the rotational period through the brightness modulation caused by large spots.

The RS CVn systems are well suited to test the tidal theories, and were indeed used this way by Hall and Henry [HH90]. There might be a problem, however, because according to e.g. Montesinos et al. [MG88], some of these systems may undergo mass-transfer, like SZ Psc (see [B81], [HR84]). In the case of SZ Psc, there seems to be direct evidence of such a transfer, since H_α emission and orbital period changes have been observed. In other systems, mass-transfer is just suspected on the basis that the most evolved component's mass is either the same as, or smaller than, the mass of its companion. In such cases, it is much more difficult to discuss the tidal history, since the mass-transfer may have modified both the orbital characteristics of the system and the rotational velocity of the components (at least of the mass-gaining star, the mass-loosing one presumably filling its Roche-lobe and being therefore synchronized).

RZ Eridani was put by Montesinos et al. [MG88] in the category of those systems with "abnormal evolution", because both components have nearly the same mass, while the cooler one has a much larger radius than the hotter one's. We show in the third section that there is no compelling reason to consider RZ Eri this way, because normal evolution is quite compatible with all observed data when allowance is made for their uncertainties, even though these have been strongly reduced since the work of Montesinos et al.

2.   Observations

The photometric observations have been made from La Silla, Chile, between November 22, 1977 and April 24, 1989. 383 measurements were made in each of the seven passbands of the Geneva system (for the definition of this system and its passbands, see [G80], [R88], and [RN88]), always with the same, two-channel aperture photometer (see [BR79]). Before 1980, a 40cm telescope was used, while a 70cm one was used later on. The results have been interpreted and published by Burki et al. [BK92]. Let us recall here a few important points:

The V magnitude of the system is about 7.75 outside eclipse, and drops to 8.9 in the primary minimum. The secondary eclipse occurs at phase 0.67 (as first suspected by Caton [C86]) and is much less deep than the primary (about $0.^{m}08$). The orbital period is

$$P = 39.28254 \pm 0.00005 \text{days}$$

and the origin of the phases (taken at primary minimum) is

$$T_o = 2446048.883 [\text{HJD}]$$

Superposed on this lightcurve, are seen periodic variations (the so-called "migration wave") whose amplitude and shape vary from one season to another. Their period also varies, but only within a very restricted range: 34.56 days to 36.25 days. They are interpreted as due to large spots on the cooler companion which rotates with the observed period, the variation of which reflecting only the differential rotation and the different latitudes of the spots. When the (constant) contribution of the hotter companion is removed, the amplitude of this "migration wave" is in the order of $0.^{m}2$. During the season October 1987 – April 1988, this variation has been especially regular, allowing a satisfactory fit to be made with a relatively simple Fourier series.

We did not measure ourselves the radial velocity curve, but relied on that published by Popper [P88].

A significant amount of *circumstellar* extinction is found from the analysis of both our ground-based photometric data and of UV spectra taken with the IUE satellite in the low dispersion mode (see [BK92] for details).

3.   Parameters and evolutionary state of the RZ Eri system

In order to obtain the photometric parameters, we had first to correct the observed magnitudes for the intrinsic variations of the cooler component (for convenience, we shall call B the cooler component and A the hotter one, following Burki et al. [BK92], although it should be the reverse according to stellar evolution). This is not quite straitforward when considering the "migration wave", because we do not know what are the respective contributions of the darks spots and of the bright plages. We have therefore analyzed the lightcurve (using Etzel's EBOP programme) for two different corrected magnitudes of the B star, assuming equal contributions of dark and bright spots in one case, and only dark spots in the other. The resulting geometrical elements are practically the same in both cases, and are listed in Table 1. The evolutionary state of RZ Eri and of three other systems can be inferred from the HR diagram shown in Fig. 1, where we have plotted the observed data together with the most recent evolutionary tracks of Schaller et al. [SS92] for Z = 0.020.

Table 1. Parameters of RZ Eri obtained through the EBOP programme from the analysis of the lightcurve.

Parameter		V band	
Primary radius (rel. to a)	r_A	0.0385	±0.0006
ratio of radii	k	2.44	±0.03
oblateness		0.00009	±0.00124
eccentricity	e	0.377	±0.008
long. of periastron	ω [°]	-47.3	±1.2
inclination	i [°]	89.31	±0.13
ratio of central surf. brightness		0.109	±0.003
normalized prim. lum.	L_A	0.627	
normalized sec. lum.	L_B	0.372	
residual stand. dev.	σ_{res} [mag.]	0.0047	

It seems that the B component is ascending the RGB, while the A component is just beginning to leave the main sequence. On Fig. 1 are also plotted three other binary systems containing at least one evolved star: AI Phe. TZ For and Capella. The characteristics of these systems have been published by Anderson et al. [AC88], [AC92].

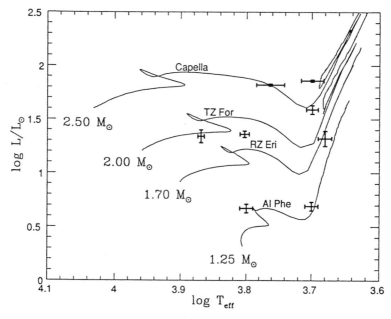

Fig. 1: HR diagram showing the evolutionary state of both components of RZ Eri and of three other, similar systems. RZ Eri B and AI Phe B are still ascending the red giant branch, while TZ For B and Capella B are probably already in the He-burning phase.

In Table 2 are listed the physical parameters of RZ Eri and of the three other systems. Unfortunately, in the case of RZ Eri, the individual masses are not known with enough precision to decide which component is the more massive. An interpolation between the evolutionary tracks of Schaller et al. [SS92] shows that, if the A (hot) component's mass is 1.70 $M_{\odot}$, then it is sufficient that the other component has a mass of about 1.73 $M_{\odot}$ to reach the observed position in the HR diagram. This is well within the uncertainty of the mass determination, so that there is as yet no evidence that any mass-transfer occured, i.e. that the components of this system have not evolved as single stars. The presence of circumstellar matter, betrayed by a significant colour excess [BK92] and by an IR excess [BS90] means that some amount of mass loss (from the B component) has taken place. However. a rough estimate shows that only about 10^{-4} $M_{\odot}$ or less of circumstellar matter are needed to explain the observations, and this is *not more* than what is expected from the parametrisation of the mass loss rate proposed by de Jager et al. [JN88] for the *normal stars* (see [BK92] for details). Thus, the suspicion of Montesinos et al. [MG88] that this system underwent abnormal evolution is not justified as yet, although existing data do not invalidate it either.

The ages mentioned in Table 2 are those inferred from the positions of the B components of the systems in the HR diagram (Fig. 1) and from models with M = 1.25, 1.7, 2.0 and 2.5 $M_{\odot}$ for AI Phe, RZ Eri, TZ For and Capella respectively. This is only a first approximation which was considered sufficient in this preliminary discussion.

Table 2. Physical parameters of the 4 systems shown in Fig. 1. The data on AI Phe are taken from Andersen et al. [AC88], those on RZ Eri from Burki et al. [BK92] and those on TZ For and Capella from Andersen et al. [AC92].

	AI Phe	RZ Eri	TZ For	Capella
Mass of A (hotter) [$M_{\odot}$]	1.195 ± 0.004	1.69 ± 0.06	1.95 ± 0.03	2.5 ± 0.2
Mass of B (cooler)	1.236 ± 0.005	1.63 ± 0.13	2.05 ± 0.06	2.6 ± 0.4
R_A [$R_{\odot}$]	1.816 ± 0.024	2.79 ± 0.12	3.96 ± 0.09	8.1 ± 0.8
R_B	2.93 ± 0.048	6.80 ± 0.23	8.32 ± 0.12	11.4 ± 1.2
P_{orb} [d]	24.59	39.28	75.67	104.02
eccentricity	0.189	0.377	0.	0.
P_{ps} [d]	20.2	21.4	75.67	104.02
$P_{rot}A$ [d]	23.1 ± 6.0	2.4 ± 0.5	4.7 ± 0.3	7.8 ± 0.9
$P_{rot}B$ [d]	24.9 ± 4.5	34.5 ± 2	105 ± 28	80 ± 35
a [$R_{\odot}$]	47.8	72.5	119.5	160.1
age [10^9 y]	5.24	1.88	1.18	0.6

4. Tidal evolution

We propose here a preliminary discussion of the tidal history of RZ Eri as well as of the three other systems. First of all, we see from Table 2 that neither RZ Eri A nor

RZ Eri B is synchronized, although the latter rotates at a *slower* speed than expected, because in the case of an elliptical orbit, the angular axial velocity of the components tends to synchronize with the orbital angular velocity *at periastron*, as shown by Hut [H80], [H81]. The corresponding "pseudosynchronous period" is obtained from equation 42 of [H81] and listed in Table 2.

Both components of AI Phe may be considered as pseudosynchronized, although their orbit is still eccentric. This is quite coherent with the theoretical expectation that synchronization is achieved much earlier than circularization (see e.g. [Z77]). TZ For and Capella are already circularized, and their cooler component is synchronized but their hotter component is *not*.

4.1.   Synchronization

In order to discuss these systems, we have first computed their synchonization time according to Zahn [Z77] for stars with a convective envelope, as a function of time. The k_2 constants were taken from the work of Claret and Gimenez [CG91], and were considered approximately valid in our case although the models used by these authors are a bit different from those of Schaller et al. [SS92], from which the moments of inertia were computed. An example of such diagrams is shown in Figure 2 for RZ Eri and TZ For. Although they are a bit rough (the formula 4.6 of [Z77] does not take the eccentricity into account), they nevertheless show a few interesting features:

– for RZ Eri B, t_{sync} decreased sharply from about 10^7 to less than 10^6 years in the last twenty million years or so, leaving well enough room for pseudosynchronization to take place, in contradiction with observations.

– RZ Eri A is not far from being synchronized, since t_{sync} is close to 10^8 years, near the end of the main sequence phase. However, this star is only beginning now to develop a deep outer convective zone, while its envelope was essentially radiative before, and t_{sync} was accordingly much longer [Z77].

– TZ For B could not be synchronized during its main sequence lifetime for two reasons: first because it had an essentially radiative envelope, second because even if it had a convective one, t_{sync} would have been greater than the time spent on the m.s. It became efficiently synchronized $\approx$ 5 10^7 years ago, when it was ascending the RGB.

– TZ For A is on the verge of being synchronized: it will be in 10^7 years or so, partly because its convection zone is rapidly growing inwards.

Similar diagrams (not shown here for the sake of brevity) clearly show that:

– both components of AI Phe must be pseudosynchronized, especially as they had a significant outer convective zone throughout their main sequence life.

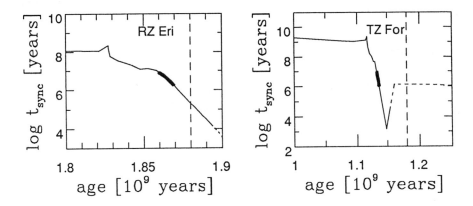

Fig. 2: Synchronization time (from [Z77]) vs. age for RZ Eri B and TZ For B. The age range where synchronism is taking place is indicated by a thicker line. The vertical dashed line indicates the present age of the system. The dashed prolongation of the continuous line indicates an extrapolation of the k_2 constant.

– Capella B was synchronized 10^7 years ago during its ascent towards the tip of the RGB, while Capella A is just beginning *now* to be synchronized.

In conclusion, these very preliminary considerations are quite consistent with observations for all systems but RZ Eri, as far as synchronism alone is concerned.

A more rigorous way of comparing theory with observations is to start from realistic initial conditions and to integrate numerically the differential equations which govern both synchronization and circularization. These equations are coupled and must be solved simustaneously. We considered only RZ Eri and AI Phe (for the two other systems, the above discussion is sufficient) and started from Zahn's [Z77] equations 4.3 to 4.5 (corrected according to [Z78]). The initial axial rotation period was fixed at one day, which appears reasonable (spectral types $\approx$ A3 to F6 on the ZAMS); in any case this parameter is far from crucial, since it is "forgotten" by the star before the end of the main sequence phase. We first assumed that the orbital parameters have remained constant in the past, which is approximately true, and slightly modified the initial semi-major axis and eccentricity by trial and error to match their present value. The quadratic term in eccentricity was taken into account. We admitted that whenever the star's effective temperature is greater than 7000 K, the envelope may be considered as radiative and the angular momentum remains practically constant, the spin frequency Ω changing only because of the varying moment of inertia. Tidal effects are supposed to turn on only when $T_{eff} < 7000$ K. Figure 3 shows the results for both RZ Eri and AI Phe. We see that the increase of angular momentum temporarily overcomes the tidal effects after the end of the main sequence phase, as already noticed by Habets and Zwaan [HZ89]. while pseudosynchronism is reached aftewards.

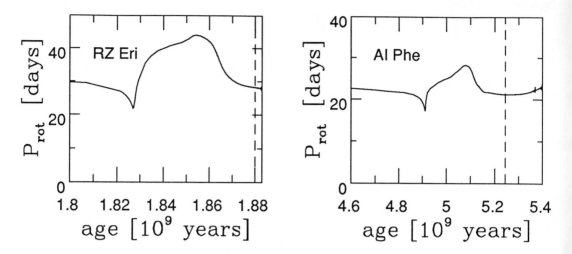

Fig. 3: Spin period (from [Z77]) vs. age for RZ Eri B and TZ For B. The vertical line indicates the age of the system. The sharp dip corresponds to the end of the main sequence phase.

The period settles at about 21 days in the case of AI Phe, which is very close to the expected pseudosynchronous period. In the case of RZ Eri, it settles around 28 days instead of the expected 21.4 days, maybe because the eccentricity is too large to be correctly taken into account by Zahn's approximations. The observations confirm the theoretical prediction for AI Phe, where both companions are pseudosynchronized (the hotter component would lie on a quite similar curve, but on the ascending branch of the dip) but not for RZ Eri, whose B component rotates *slower* then the pseudosynchronous frequency, and should now be *accelerated* by the tide. According to Fig. 3, one would even expect RZ Eri A to be rather close to (pseudo)synchronism, again in contradiction with the observations, but this prediction is too sensitive to the somewhat arbitrary limit between "convective" and "radiative" envelopes. The observed fact, that RZ Eri B rotates slower than pseudosynchronism does not necessarily invalidates Zahn's 1977 formulae (which anyway are rather approximate compared with the improved theory [Z89]), because magnetic braking is not included in our calculation. To take it into account is not a simple matter, especially as the amount of mass loss is unknown. In this context, a very precise measurement of the radial velocity curve of each companion would be extremely valuable, provided it allows a tenfold improvement (at least!) of the precision of the individual masses: a detailed comparison with models may then tell whether the mass lost by the B component has been high enough to significantly alter its nuclear as well as its rotational (through magnetic braking) evolution.

4.2. Circularization

The circularization time has been computed from the theory of Zahn [Z77] as well as from the theory of Tassoul [T88], and plotted as a function of age for the four systems. It turns out that:

– AI Phe is not circularized yet according to Zahn's theory (although it will be shortly), while it should be according to Tassoul's theory. Since the observed eccentricity is significant, Zahn's theory seems better in this particular case.

– RZ Eri should begin to circularize now, or be already circularized. Here also, Tassoul's mechanism seems too efficient.

– TZ For and Capella have been circularized $4 \ 10^7$ and $5 \ 10^6$ years ago respectively, when the B component was reaching the tip of the RGB and *had a much larger radius then now*. This is predicted equally well by both theories, and completely confirm the discussion of Andersen et al. [AC92], as shown in Figure 4.

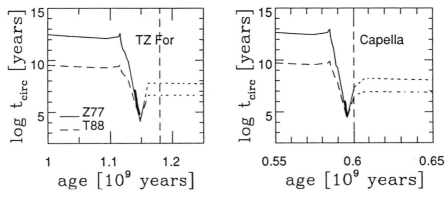

Fig. 4: Circularization time (from [Z77] and [T88] vs. age for TZ For and Capella. Same codification as in Fig. 2. The narrow dip corresponds to the tip of the red giant branch, where the radius of the B component reaches a maximum.

Integration of the differential equations shows that the eccentricity of the RZ Eri system is just beginning to decrease (Figure 5), and an extrapolation of the tendency indicates that circularization will be reached within 10^7 years or so, thanks to the increasing radius of the B component ascending the giant branch. The same is true of AI Phe, although the orbit will not be circular before about $2 \ 10^8$ years, because of the slower evolution of the B component.

5. Conclusion

This work (an extension of [BK92]) has outlined the interest that RZ Eridani may have in the context of tidal evolution. Its interest comes from the intermediate situa-

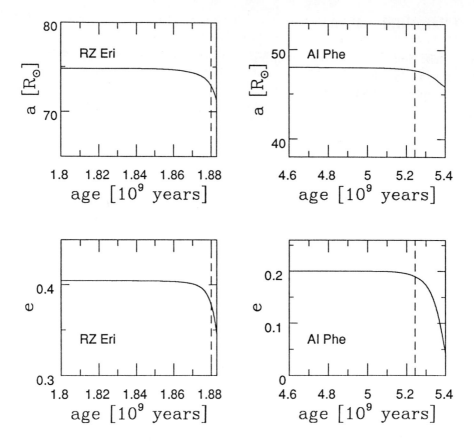

Fig. 5: Semi-major axis (top) and eccentricity (bottom) vs. time for RZ Eri (left) and AI Phe (right). The vertical line indicates the age of the system, where a and e reach the observed values.

tion of this system between binaries like AI Phe, which are synchronized but not yet circularized, and binaries like TZ For whose cooler component alone is synchronized and which are circularized. Furthermore, the tidal characteristics of RZ Eri are not exactly reproduced by the tidal theories used here. This may be caused by the use of an inadequate model of stellar structure (inexact mass, chemical composition and possibly mass loss), by the roughness of the tidal theory (a more elaborate theory, like e.g. [Z89], should be used in a more detailed discussion), and by the neglect of magnetic braking. But in any case, this discrepancy implies that RZ Eri may become an interesting constraint of the tidal theories, especially when the individual masses will have been determined with a much improved precision. A first step is being made by Imbert (private communication) through Coravel observations which will constrain the mass of the cooler component. Classical spectroscopy is needed to obtain a precise radial velocity curve of the hotter component, which remains out of the range of Coravel because of its temperature and rotational velocity.

Acknowledgements: We thank our colleagues F. Rufener, N. Cramer and C. Richard for the reduction of the photometric maesurements of RZ Eri, and M. Burnet for the technical maintenance of the Geneva station at La Silla during the whole survey. We are also indebted to Dr. P.B. Etzel for providing us the EBOP code, and to Dr. G. Meynet for his help in computing the moment of inertia of his models. This work was supported in part by the Swiss National Foundation for Scientific Research.

References

[AC88]   Andersen J., Clausen J.V., Gustafsson B., Nordström B., VandenBerg D.A., 1988, A&A 196, 128

[AC92]   Andersen J., Clausen J.V., Nordström B., Tomkin J., Mayor M., 1992, A&A 246, 99

[BK92]   Burki G., Kvíz Z., North P., 1992, A&A (in press)

[B81]   Bopp B.W. 1981, AJ 86, 771

[BR79]   Burnet M., Rufener F., 1979, A&A 74, 54

[BS90]   Busso M., Scaltriti F., Ferrari-Toniolo M., Origlia L., Persi P., Robberto M., Silvestro G., 1990, Mem. Soc. Astron. Ital. 61, 77

[C86]   Caton D.B., 1986, AJ 91, 132

[CG91]   Claret A., Giménez A., 1991, A&AS 87, 507

[FMH86] Fekel F.C., Moffet T.J., Henry G.W., 1986, ApJS 60, 551

[G80]   Golay M., 1980, Vistas in Astronomy 24, 141

[HZ89]   Habets G.M.H.J., Zwaan C., 1989, A&A 211, 56

[HH90]   Hall D.S., Henry G.W., 1990, in: Ibanoglu C. (ed.) *Active Close Binaries. Kluwer Academic Publishers, the Netherlands, p. 287.*

[HR84]   Huenemoerder D.P., Ramsey L.W., 1984, AJ 89, 549

[H80]   Hut P., 1980, A&A 92, 167

[H81]   Hut P., 1981, A&A 99, 126

[JN88]   de Jager C., Niewenhuijzen H., van der Hucht K.A., 1988, A&AS 72, 259

[M52]   Motz L., 1952, ApJ 115, 562

[MM84]   Mayor M., Mermilliod J.-C., 1984, in: Maeder A., Renzini A. (eds.) *IAU Symposium 105, Observational Tests of the Stellar Evolution Theory*. Reidel, Dordrecht, p. 411

[MG88]   Montesinos B., Giménez A., Fernández-Figueroa M.J., 1988, MNRAS 232, 361

[P88]   Popper D.M. 1988, AJ 96, 1040

[R88]   Rufener F. 1988, *Catalogue of Stars Measured in the Geneva Observatory Photometric System (4th edition)*. Observatoire de Genève

[RN88]   Rufener F., Nicolet B., 1988, A&A 206, 357

[RP88]   Rutten R.G.M., Pylyser E., 1988, A&A 191, 227

[SS92]   Schaller G., Schaerer D., Meynet G., Maeder A., 1992, A&AS (in press)

[T87]   Tassoul J.-L., 1987, ApJ 322, 856

[T88]   Tassoul J.-L., 1988, ApJ 324, L71

[Z77]   Zahn J.-P., 1977, A&A 57, 383

[Z78]   Zahn J.-P., 1978, A&A 67, 162

[Z89]   Zahn J.-P., 1989, A&A 220, 112

Ekman layers and tidal synchronization of binary stars

Michel Rieutord *

Abstract

Ekman circulation has recently been proposed by Tassoul to be a mechanism by which close binaries get synchronized. We have shown (see [Rie92]) that such a mechanism is in fact inefficient to drive the synchronization because of the stress-free boundary conditions met by the fluid at the surface of the star.

Reference

[Rie92] Rieutord M. *Ekman circulation and the synchronization of binary stars. 1992 Astron. Astrophys. (in press)*

*Observatoire Midi-Pyrénées, 14 avenue Edouard Belin, 31400 Toulouse, France.

The Distribution of Mass Ratio in Late-type Main-sequence Binary Systems

Andrei A. Tokovinin *

Abstract

The maximum likelihood method used to derive the distribution of mass ratio q from the combined data on single and double-lined binaries is described. It's validity is demonstrated by numerical simulations. The results of the radial velocity survey of nearby K and M dwarfs are presented, the extended KM-dwarf sample and the sample of nearby G-dwarfs are analyzed as well. The distributions of q in these 3 samples are flat and the differences between them appear to be not significant. However the frequency of spectroscopic binaries with periods less than 3000 days among K and M dwarfs is lower than for G dwarfs and the cutoff in the secondary mass distribution at 0.08 solar mass is seen.

1.   Introduction.

The radial velocity survey of a well chosen sample of stars can provide answers to the following questions:

– What is the frequency of short-period binaries?

– What is the shape of the mass ratio distribution?

– Are there any substellar mass secondary companions?

In the analysis of the data provided by such a survey it is necessary to consider various selection effects and the influence of random orbital inclinations. The data analysis scheme based on the maximum likelihood principle will be explained first. Then three samples of nearby dwarf stars will de described and the distributions of the mass ratio q in these samples will be discussed.

---

*Sternberg Astronomical Institute, 119899 Universitetsky prosp., 13 Moscow, Russia. Internet: SNN@SAI.MSK.SU (shared)

2. Derivation of mass ratio distribution.

The best estimate of the distribution of the mass ratio $P_q(q)$ is found by maximizing the probability L of given observations. It is called the likelihood function and it is equal to the product of the probabilities P_i of the observations of i-th star. All information contained in the observations is eventually expressed by these quantities P_i.

Radial velocity observations of a homogeneous sample of stars carried out over a sufficiently long time lead to the discovery of a number of binary systems and, eventually, to the determinations of their orbits. It is natural to place un upper limit to the orbital periods taken into consideration in order to assure reasonable detection limits. For the single-lined binaries (SB1) only mass functions are known. If the mass of the primary is estimated from its color the mass function can be replaced by the equivalent quantity $F = qx(1+q)^{-2/3}$ where $x = \sin i$. For SB2 the mass ratio q is measured directly.

The combined analysis of SB1 and SB2 data requires some caution. The probability of non-detection of the lines of secondary components depends on the mass ratio q and is modelled by the function $d(q)$ which goes from 1 at $q = 0$ to 0 at $q = 1$. The probabilities D_{SB1} and D_{SB2} for a spectroscopic binary to be single- or double-lined are expressed as integrals over q of mass ratio distribution P_q multiplied by d or $1-d$.

We have chosen to represent the transition from non-detection to complete detection by the linear drop within the q-bin that contains the lowest-q SB2 in a given sample. The numeric simulations show that incorrect modelling of SB2/SB1 partition produces in the restored distributions maxima and minima at the SB1/SB2 margin. Since the real process of secondary line detection can hardly be described by any simplistic model the features of q distributions around $q = 0.7$ should be regarded with suspicion. It is of course possible to treat all binaries as SB1 but this would inevitably entail some loss of information.

The distribution of mass ratios in the SB1 sub-sample is equal to $P_q(q)d(q)$. The distribution of the mass-function equivalent F is obtained by its convolution with the distribution of $x = \sin i$ that results from the hypothesis of random orientation of orbital planes:

$$P_F(F) = \frac{1}{F} \int_{q_{min}}^{1} \frac{x^2}{\sqrt{1-x^2}} P_q(q)d(q)\mathrm{d}q. \tag{1}$$

At any given F there is a relation between x and q. The lower limit of integration q_{min} corresponds to $x = 1$. The singularity is integrable and for practical purposes it is more convenient to carry the integration over x rather than over q.

The observation of i-th SB1 consists of two distinct facts: i) double lines are not seen and ii) the mass function is equal to F_i. The probability of observation P_i is equal to $P_F(F_i) \cdot D_{SB1}$. Similarly for SB2 it is equal to $(1 - d(q_i)) \cdot P_q(q_i) \cdot D_{SB2}$. Here the first multiplier can be safely omitted since it does not depend on the q-distribution. The normalization of P_i is of no importance. For the same reason the use of mass function or any other related quantity instead of F for SB1 would lead to the same result.

In practice the distribution is discretized into K bins, $P_q(q) = f_k$ for $q_{k-1} < q < q_k$. The probabilities P_i are then represented as scalar products, $P_i = \sum_k p_{ik} f_k$, and the natural logarithm of the likelihood function is

$$\ln L = \sum_{i=1}^{2N} \ln \sum_{k=1}^{K} p_{ik} f_k, \qquad (2)$$

where N is the total number of spectroscopic binaries in the sample. The elements of the first N rows of the matrix $\{p_{ik}\}$ are easily calculated using (1) for SB1 and are equal to 0 or 1 for SB2. The elements of the next N rows come from the detection probabilities D_{SB1} or D_{SB2} and are equal to the integrals of d or $1 - d$ over k-th bin.

Each term in (2) should be divided by the probability of the detection of binarity to take into account undetected systems. This was not done here since detection probabilities for SB1 are always close to 1. The detection of SB1 was studied in (Tokovinin 1992 – hereafter [T92]) by numerical modelling. Virtually all SB2s with periods less than 3000 days are detected as well.

There exists a very efficient iterative technique for finding the maximum of (2) over $\{f_k\}$. It was described by Lucy [L74]. He also applied it to the correction of the observed distribution for the effects of random inclinations and noted its usefulness in a more general context of the solution of inverse problems. The iterative method of Mazeh and Goldberg [MaG92] for finding the distribution of q is similar to that of Lucy.

Inverse problems are well-known to be very sensitive to the noise in the input data. Our numeric simulations indeed show that the maximum likelihood solution is quite satisfactory for big samples (more than 50 stars per bin) but tend to have irregular maxima and minima when the sample is small. A similar behavior was noted in tomographic image restoration (Llacer and Nuñez [LlN90]).

In the real life the size of sample is usually small and some regularization prescription is necessary to produce a "good-looking" distribution. Lucy suggested to stop iterations well before the convergence. When the uniform distribution is taken at the first step this prescription gives a regularized solution which is biased towards uniform distribution.

A somewhat better approach seems to be the search of the most smooth distribution that is compatible with the observations. The natural degree of compatibility is offered by the likelihood function itself. If $\{f\}$ is the exact solution then any distribution $\{g\}$ is plausible if $\ln L(f) - \ln L(g) < 1$ since it means that $\{g\}$ would make the probability of our observations only e times smaller. The choice of 1 as a plausibility threshold is, of course, arbitrary.

We use the sum of $(f_k - f_{k-1})^2$ as the smoothness criterion and add it with some negative coefficient to $\ln L$. This coefficient is chosen by trial and error to give a solution with $\ln L$ that is smaller by 1 than its maximum value. Thus a distribution that is biased towards maximum smoothness but still compatible with the observations is obtained.

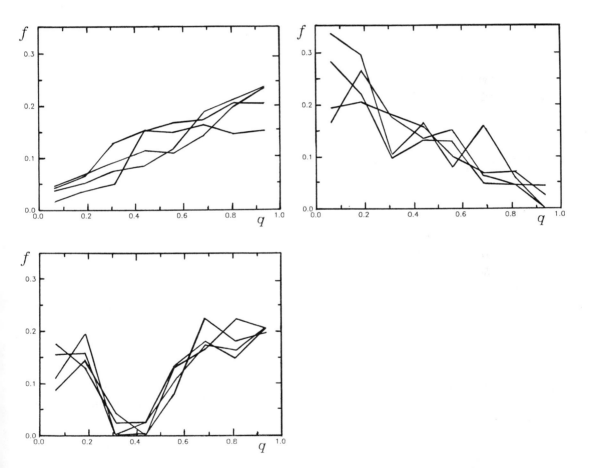

Figure 1. Results of the numeric simulations. Samples of 50 artificial binary stars with known distribution of q were generated and analysed: (a) linear growing distribution, (b) linear falling distribution,   (c) uniform distribution with a gap at $q = 0.25 - 0.5$. Results for 4 realizations are plotted.

The results of numerical testing of the data analysis scheme described above are given in Fig. 1. The samples of 50 artificial binary stars with known q distributions were generated and then analyzed just as real data. Not only are the linearly rising and falling distributions well restored but the distribution with a sharp gap is also reasonably well recovered despite the fact that it is not at all smooth.

3.   Radial velocity surveys of dwarf stars.

Our radial velocity survey of K and M dwarfs in the solar neighbourhood was started in 1985. The sample included stars from the Gliese catalogue [Gli69] and its supplement [GlJ79] with visual magnitude brighter than 10^m, $B - V$ color from 0.76 to 1.5 (spectral types from G9 to M3). Only limited part of the sky with $\delta > -10°$ and α in $14^h..0^h..5^h$ zone was surveyed. Systems with blended visual companions and subdwarfs were excluded. The total number of stars is 200.

The correlation radial-velocity meter (Tokovinin [Tok87]; Goryatchev et al. [GKS88]) was used to measure radial velocities with the typical precision of 0.5 km/s. The limiting accuracy of this instrument is about 0.3 km/s. Velocity zero point was referenced to the IAU standards. Several telescopes located at the different observatories (Moscow, Crimea, Maidanak, Abastumani) were used. On the average 7 observations per star have been made. More information on this survey and its results can be found in [T92]. The first 4 orbits are published in [Tok91].

The KM-dwarf sample contains 10 single-lined and 12 double-lined binaries with periods less than 3000 days. However 5 of SB2's are just at the magnitude limit of the sample and would not be included were they single stars. They were removed from further analysis.

The extended KM-dwarf sample is obtained by adding to all our data the results of recent radial velocity surveys: those of Hyades by Griffin et al. [GGZ85], a subdwarf survey of Latham et al. [LMC88] and some orbits published by Duquennoy and Mayor [DuM88a], [DuM88b]. Only stars with $B - V > 0.7$ and periods greater than 3000 days were selected.

The sample of G-dwarfs observed by Duquennoy and Mayor [DM91] with periods less than 3000 days was also analyzed. The general characteristics of these 3 samples are summarized in Table 1.

4.   Results and discussion.

The restored distributions of mass ratio are shown on Fig. 2. The features around $q = 0.7$ are most likely not real but caused by the SB1/SB2 dichotomy. In all 3 samples the distribution of mass ratio looks surprisingly flat. The deficiency of binaries with the smallest q reflects the absence of substellar mass companions. A

Table 1.

Sample	Number of SB1	Number of SB2	Lowest q
Main KM	10	7	0.65
Extended KM	32	25	0.65
G-dwarfs	19	13	0.55

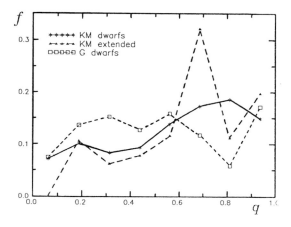

Figure 2. Distribution of mass ratio q in three samples of nearby dwarf binary stars with periods less than 3000 days.

more thorough analysis of their detection limits and the upper limits of their frequency are given in [T92]. The rise towards $q = 1$ in the extended KM sample can be explained by its heterogeneous nature and hence the influence of selection effects in favor of SB2.

It is now clear that the gap in the mass distribution of the secondary components at 0.2-0.3 solar mass that was "discovered" in [T92] is due to the incorrect treatment of SB1/SB2 selection effects together with the absence of any regularization.

The data on single stars must be included in the analysis if the estimation of binary frequency is needed. This was done in [T92] and will not be repeated here. It appears that the frequency of spectroscopic components with periods less than 3000 days for KM dwarfs ($10 \pm 2\%$) is less than for G dwarfs studied by [DM91] that is estimated to be 17% from their Fig.7. This fact may be related to the existence of lower limit of companion mass.

Acknowledgements. The author is grateful to M.Mayor and A.Duquennoy for useful and stimulating discussions.

References

[DuM88a] Duquennoy A., Mayor M. *1988a, A&A, 195, 129*

[DuM88b] Duquennoy A., Mayor M. *1988b, A&A, 200, 135*

[DM91] Duquennoy A., Mayor M. *1991, A&A, 248, 485*

[Gli69] Gliese W. *1969, Veröff. Astron. Rechnen Inst. Heidelberg N.22, Catalogue of Nearby Stars*

[GlJ79] Gliese W., Jahreiss H. *1979, A&AS, 38, 423*

[GKS88] Goryachev M.V., Klementyeva A.Yu., Semenikin A.A., Tokovinin A.A. *1988, Trudy GAIS (Proc. of the Sternberg Astron. Inst.) 60, 242*

[GGZ85] Griffin R.F, Gunn J.E., Zimmerman B., Griffin R.E.M. *1985, AJ, 90, 609*

[LMC88] Latham D.W., Mazeh T., Carney B.W., McCrosky R.E., Stephanik R.P., Davis R.J. *1988, AJ, 96, 567*

[LlN90] Llacer J., Nuñez J. *Iterative maximum likelihood estimator and bayesian algorithms for image reconstruction in astronomy. In: The restoration of HST images and spectra. Ed. R.L.White, R.J.Allen. StScI, Baltimore, 1990*

[L74] Lucy L.B. *1974, AJ, 79, 745*

[MaG92] Mazeh T., Goldberg D. *1992, These Proceedings*

[Tok87] Tokovinin A.A. *1987, AZh, 63, 196 (Sov. Astron., 31, 98)*

[Tok91] Tokovinin A.A. *1991 A&AS, 91, 497*

[T92] Tokovinin A.A. *1992, A&A, in print*

The Dynamical Evolution of G-type Main Sequence binaries

Frans van 't Veer [*]   *Carla Maceroni* [†]

Abstract

We will speak about the dynamical evolution of AMLOSC binaries.
AMLOSC binaries are defined as angular momentum losing binaries with orbit-spin coupling. In this paper we limit ourselves to the main-sequence group. The AML is produced by stellar winds trapped by magnetic fields. The OSC may have different origins. We mainly consider the mechanism of turbulent friction in the convective zone without neglecting however alternative solutions.
The components of AMLOSC binaries continuously approach each other until contact is reached. The dynamical evolution of the rotation of the components and the of the orbital revolution is given by Eqs.(9) and (10).
A comparison between observations and theory is given in the discussion. In the conclusions we point out, among other things, that it has no sense to determine synchronization time scales for late-type binaries from tidal interaction alone.

1.   Introduction

Low mass stars with a convective outer layer are characterized by substantial angular momentum loss (AML) carried away by stellar winds trapped by magnetic fields. Binary systems with angular momentum (AM) losing components are considered in the present paper as formed by two rotating single stars each losing AM according to its physical and dynamical parameters, independently of its close companion. Thus the problem of AML is the same as for single stars, but the big difference, for close binary systems, is the tidal interaction which is at the origin of transfer of AM (AMT) from the orbit to the components, when they are rotating slower than the orbital angular velocity ω_k and conversely.

The result of this AMT is that the AML experienced by the components is supplied by the orbital AM reservoir. Consequently the orbit will become narrower, so that,when the synchronization is good, the components will be spun up in spite of

---

[*]Institut d'Astrophysique de Paris, F-75014 Paris, France
[†]Osservatorio Astronomico di Roma, I-00044 Monteporzio Catone (RM), Italy

their AML. This apparently paradoxical situation of the increasing rotational velocity is the reason why the AML, which depends on the rotation, will in its turn increase, thus creating a process with a positive feedback: both the rotation and the approach of the components will be accelerated until the contact stage is reached.

During the contact stage the AML will continue, and mass transfer from the secondary to the primary will take place. Finally the binary the two components will coalesce into a single star.

Evidently the tidal interaction depends on the closeness, and hence the orbital period P_K of the components. When the period is too large the AMT becomes negligible and the components will spin-down like single stars without change of the orbit.

The origin of our knowledge of rotational braking by stellar winds trapped by magnetic fields, goes back to a paper by Schatzman (1962) [Sch62]. Since that time it was recognized that magnetic fields also play an important role for the dynamical evolution of close late-type binary stars (Huang [Hua66], Okamoto and Saito [OkS70], van 't Veer [VtV79], Rahunen [Rah81], Vilhu [Vil82], Eggleton [Egg82], Mestel [Mes84]). This dynamical evolution also strongly depends on the tidal interaction between the two components, whose theory was independently developed in the same period (see next paragraph).

In the present paper we will review the dynamical evolution of close solar type binaries as it is controlled by the AML mechanism. We will not consider AML questions related to the late stages of stellar evolution.

In the next paragraph we present a set of differential equations from which the dynamical evolution can be derived. In the third paragraph we compare the results derived from the theoretical formulation with the observational data. In the final discussion and conclusions we will examine the weak points of both theoretical and observational knowledge of AML.

2.   The angular momentum transport in a binary

For an AM losing binary with tidal coupling between orbit and spin of the components (AMLOSC binary) we have to consider the AML from the components and the AM transfer (AMT) between the orbit and the spin of the components.

Angular momentum loss (AML). The partly ionized stellar wind, from the components, trapped by corotating magnetic fields are at the origin of the AML of the binary system. The distance of corotation of the outflowing particles, or Alfven radius (R_A), depends on the geometry of the magnetic field which in its turn depends on the velocity of rotation and on the thickness of the convective zone. Only theoretical deductions and speculations are available for the structure of the magnetic field, which can be observed only in the case of the slowly rotating sun. The absence of knowledge concerning this structure and the subsequent need of ad hoc hypotheses should be considered as one of the weak points of any model.

The resulting stellar AML can be measured:
1. from a comparison of the rotational velocity of stars of different ages or
2. from the period distribution of close solar type binaries.
In the first case we suppose that all G-type MS stars follow, after their formation, a similar rotational deceleration. In this way one can derive an acceptable general spin-down relation which is directly transformed into a general law for AML. The second case will be discussed below at the end of this paragraph.

Even if this comparison of different stars of different ages should hold reasonably well we must not forget that in reality we do not measure the AML, but just the spin-down of the visible outer layers. So we must convert the change of rotation into AML.

For an isotropic stellar wind we often use

$$\dot{H} = \frac{2}{3}\dot{m}R_A^2\omega \tag{1}$$

relating directly AML to mass loss (ML) by means of the Alfven radius. For a homogeneous rotator losing AM and neglecting the corresponding ML and change of radius we may write

$$H = m(kr)^2\omega \tag{2}$$

giving

$$\dot{H} = m(kr)^2\dot{\omega} \tag{3}$$

where kr is the radius of gyration. This formulation supposes that the AML is acting directly on all the layers of the star. In reality we should expect a direct action only on the convective zone combined with a delayed action on the inner layers, depending on the coupling of the two parts.

We do not possess observational data or sufficient theoretical insight concerning this last question, and moreover there are no reasons to suppose that stars are rotating in concentric layers as it would have been nice for an easier formulation. Therefore, we introduce a parameter $x \leq 1$ to take into account the importance and depth of the convective zone (CZ) and its coupling with the inner layers:

$$\dot{H} = xm(kr)^2\dot{\omega}, \tag{4}$$

From Eqs.(1) and (4) we obtain

$$d\ln\omega = \frac{2}{3}\frac{R_A^2}{x(kr)^2}d\ln m. \tag{5}$$

The high value of the ratio $R_A^2/(kr)^2$ (several thousands for the sun) is an a posteriori justification of the neglection of $\dot{m}$ and $\dot{r}$ in the differentiation of Eq.(2).

The AML given in Eq.(4) is directly related to the spin-down $\dot{\omega}$ that can be measured for single stars of galactic clusters of known, but different, age. The quantity $\dot{\omega}$ depends on the rotation ω of the star. This dependence was studied by Skumanich [Sku72], Smith [Smi79] and Soderblom [Sod83] for slow rotating stars. They all find a power law of the type

$$\dot{\omega} = c_3 \omega^{c_4} \qquad (6)$$

where c_3 is a constant and $c_4 \simeq 3$ is the exponent first defined by Skumanich [Sku72]. Its value has often been discussed theoretically (see Mestel [Mes84], Mestel and Spruit [MeS87], Vilhu and Moss [ViM86], Van 't Veer and Maceroni [VeM89]), but its empirical value depends rather crucially on the age adopted for the clusters. This so-called Skumanich relation does not hold for fast rotating stars ($P \leq 2d$). This can be seen when we examine the data for very young clusters [Sta91]. In some 10 My the rapid G-type rotators are spun down, suggesting an AML much higher than can be obtained from an extrapolation of the Skumanich relation (Eq.(6)).

From a study of the period distribution of close late-type MS binaries Maceroni and Van 't Veer [MaV91] could estimate the amount of AML taking place from the rapidly rotating components. This is the point two mentioned at the beginning of this paragraph, however for a better understanding of this question we first need to discuss the problem of AMT in a binary.

Angular momentum transfer (AMT) The process of AMT between the orbit and AM losing components has been formulated by Maceroni and Van 't Veer [MaV91], so we just present here the essential equations necessary for the understanding of the problem, supplemented by a formulation of the special simplified case involving perfect synchronization ($f = 1$).

The total AM contained in a binary is the sum of the orbital AM and the spin of the components. For equal components rotating in a circular orbit it may be written as:

$$H_{tot} = \beta' \omega_K^{-1/3} + 2xm(kr)^2 f\omega_K, \qquad (7)$$

where $\beta' = 2^{-1/3} m^{5/3} G^{2/3}$, ω_K is the orbital angular velocity, m the mass of the components and $f = \omega/\omega_K$ the degree of synchronization. From the works of Zahn [Zah77] [Zah89], Campbell and Papaloizou [CaP83] and Savonije and Papaloizou [SaP85] a general formulation of the orbit-spin action of AMLOSC binaries may be derived from the synchronization time-scale of fully convective stars. The resulting equation may be written as follows:

$$\dot{\omega} = b(\omega_K - \omega)\omega_K^\alpha, \qquad (8)$$

where $\alpha = 4$ according to most authors using Darwin's tidal torque theory (see however [Tas88]) and neglecting the interaction induced by magnetic fields. The constant b depends on the mass, luminosity, radius and the internal density distribution of the components. Its value is also very strongly influenced by the mass ratio q

(see [MaV91]). From Eqs.(6), (7) and (8) it is possible to derive the following set of differential equations:

$$\dot{\omega}_K = K x (\omega_K - \omega) \omega_K^{\alpha + 4/3} \qquad (9)$$

with $K = 6m(kr)^2 b/\beta'$ and

$$\dot{\omega} = \dot{\omega}(\omega) + b(\omega_K - \omega)\omega_K^{\alpha} \qquad (10)$$

where $\dot{\omega}(\omega)$ is the spin-down equation already discussed for slow rotating stars in Eq.(6). The solution of Eqs.(9) and (10) gives us the orbital period and the rotation of the components as a function of time. The function $P_K(t) = 2\pi/\omega_K(t)$ is called the period evolution function (PEF). It gives the AML controlled period variation provided all the parameters are known. This is indeed reasonably well the case for the constants α, b and K, but less certain for x (the coupling between convective zone and outer layers) and even badly known for the spin-down equation $\dot{\omega}(\omega)$ of fast rotating stars.

At this stage little can be said about $\dot{\omega}(\omega)$ for fast rotators, but we still have the possibility to look at the influence of the PEF on the presently observed period distribution (PPD) observed for the AMLOSC binaries. It is expected indeed that the AML will modify the initial period distribution which is defined as the period distribution at the moment of formation. This modification is limited to binary components whose spectral types are late enough to develop a convective envelope and hence a dynamo involving magnetic fields and mass loss. It also strongly depends on the strength of the orbit-spin coupling which determines the small period interval between binaries with AMT (and hence decreasing period) and binaries with negligible AMT and hence stable periods.

From the period distribution of MS binaries in the period range 0.25 to 100d we could derive ([VeM88], [VeM89], [MaV91]) the spin-down relation for rapidly rotating stars and show that it is not possible to use a unique power law formulation for slow and rapid rotators. It was necessary to admit not only a different power law but also a sudden increase of the loss for fast rotators, in order to explain the bimodal character of the period distribution, with a minimum for $P_K = 2d$ for all late-type MS binaries possessing a convective zone. The same minimum is clearly visible in the histograms of all the different groups of late-type MS binaries published by Farinella et al [FLM79] (see also [Duq92]). The impossibility to define a unique relation is confirmed by the measurements of rotational velocities of young cluster members of the same spectral type [Sta91].

In the limiting case of perfect synchronization ($f = \omega/\omega_K \rightarrow 1$) we no longer need the system of Eqs. (9) and (10), as we can simply write that the total AML is instantaneously subtracted from the orbital AM reservoir:

$$2x m (kr)^2 \dot{\omega}(\omega) = -\frac{1}{3}\beta' \omega_K^{-4/3} \dot{\omega}_K \qquad (11)$$

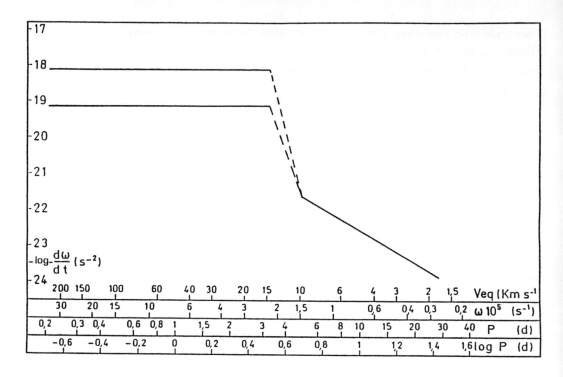

Fig. 1: Two versions for the spin-down relation $\dot{\omega}(\omega)$ introduced in Eq.(6). The right side ($v_{eq} \leq 10 \mathrm{km/s}$) represents the Skumanich [Sku72] relation with exponent $c_4 = 3$.

The left side or fast rotator part follows the choice $c_4 = 0$ derived in [MaV91] from the period statistics of close binaries. The solutions were calculated for two different coupling parameters $x = 0.1$ and $x = 1$ and give approximately the same result when used with the corresponding value of c_3 in Eqs.(9) or (12) (see text). The choice $c_4 = 0$ is approximative and should be interpreted as close to zero (see [MaV91]).

A coupling parameter of $x \simeq 1$ would give the best fit with the rotation of late-type stars in very young clusters. This value also slightly depends on the mean density distribution that is adopted for these fast to very fast rotating G0 to G9 components sitting in the MS band (radius of gyration).

This gives a new relation, of course independent from the synchronization mechanism, provided that it is considered perfectly efficient. As it supposes an extremely strong coupling of tidal or other origin it is only valid in the cases of close binaries where this coupling indeed exists. So we may write for the derivative, assuming $\omega \simeq \omega_K$

$$\dot{\omega}_K \simeq \dot{\omega} \simeq -(6x/\beta')m(kr)^2\dot{\omega}(\omega)\omega^{4/3} \tag{12}$$

as a good approximation for the complete interval of periods when the coupling is strong. Equation (12) makes it possible to obtain the PEF by direct integration, provided that the spin-down relation can be written according to Eq.(6). From the analysis of the frequency distribution of close late-type MS binaries we know however that we are forced to admit that the spin-down relation is a non regular function of ω. It is shown in Fig.1, which is a new version of Fig.1 of [MaV91], how the spin-down relation can be decomposed for the different intervals of rotation. Each interval is characterized by a spin-down relation of the power law type given in Eq.(6) with different coefficients c_3 and c_4. In the discussion we will try to see if there are physical reasons that can help us to understand this behaviour.

Equation (12) clearly shows us, in the limit of the approximation, that the same PEF can be produced by changing the coupling coefficient x as well as the coefficient c_3 of the spin-down function $\dot{\omega}(\omega)$. When the coupling x is high, the same PEF can be obtained with a smaller c_3 value than when the coupling is loose. This result is easy to understand, for when x is small the same AML will be more efficient in its effect on the observable layers. We should not forget that $\dot{\omega}_K$ is the expression of a real transfer of AM directly connected with the approach of the two components. The function $\dot{\omega}(\omega)$ however is the spin-down relation of the visible outer layers. The product $x\dot{\omega}(\omega)$ can therefore be considered as the characteristic function for the AML-controlled evolution of the binaries. Finally it is interesting to note that the factor ω_K^α which likewise produces a dependence of the period change on the period itself no longer appears in the formulation of Eq.(12).

Figures 2,3,4 illustrate the different results concerning AML, PEF and synchronization treated in this paragraph. The computations displayed in the figures have all been obtained by numerical integration of the system of Eqs. (9) and (10) (see [MaV91] for the details of the method), however we have also checked that the approximate solution presented here gives a very good approximation of the period evolution when the highly efficient synchronization mechanisms of Tassoul is assumed.

3.   Discussion

The numerical integration of the system of ordinary differential equations (9) and (10) give the solutions for the PEF without approximations. These solutions depend on the spin-down relation and on the values of α and b which, in turn, depend on the theory of tidal interaction. According to the present literature we have at our disposal two different theoretical developments for the circularization of a binary orbit.

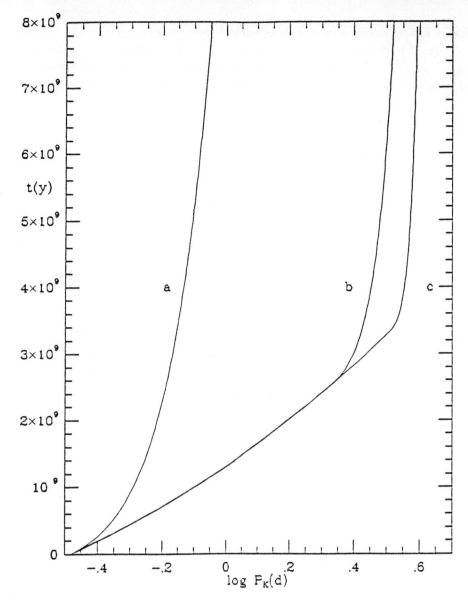

Fig. 2: The period evolution function (PEF) for 3 different cases representing 3 different combinations of the spin-down relation (AML) and tidal interaction (AMT). **a).** Skumanich relation ($c_4 = 3$) extrapolated for fast rotators and tidal interaction according to Zahn [Zah77]. **b).** Spin-down relation as derived in this paper (Fig.1) with Zahn's tidal interaction. **c).** Spin-down as for b combined with Tassoul's [Tas88] equation for interaction. With solution a we can reproduce neither the period distribution of close late-type MS binaries, nor the velocity distribution of late-type young cluster members.

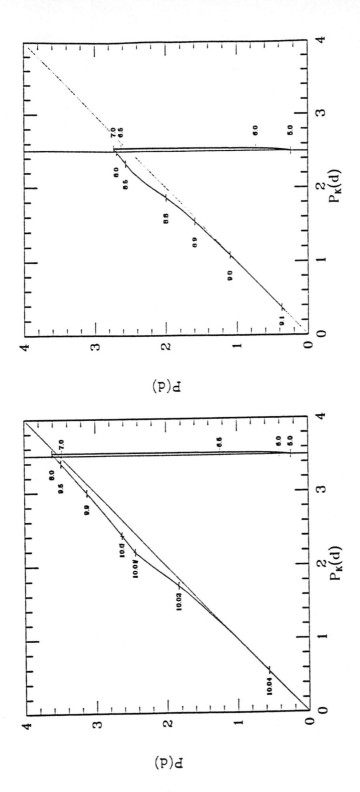

Fig. 3: The rotational evolution of AMLOSC binaries. The time steps are expressed in $\log t(y)$. The diagonal line represents the synchronization $f = 1$. Both figures are computed with the same parameters as for case b of Fig.2, they only differ for the initial value of the orbital period.

246

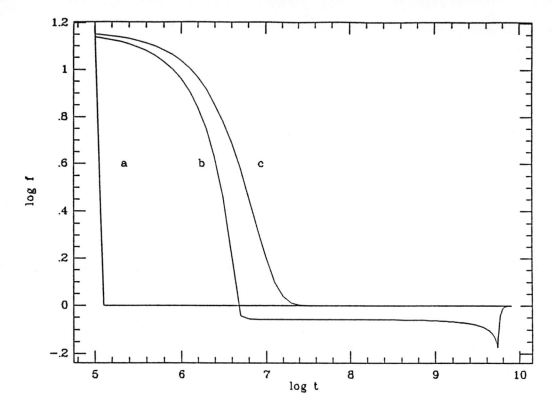

Fig. 4: The evolution of the synchronization parameter f for 3 different cases of AML and AMT.

a. Spin-down (this paper) + strong tidal coupling between the components based on Ekman advection [Tas88].

b. Same spin-down + tidal coupling based on friction produced by turbulent viscosity (see text).

c. No AML + tidal coupling as for case b.

The interesting characteristics are:

1. The difference of synchronization time-scale between a and c.

2. The absence of good synchronization for the case b. The dip at $\log t = 9.7$ is related to the jump in the spin-down relation (see Fig.1). After this dip the synchronization rapidly becomes very good.

The first one, developed among others by Zahn [Zah77], [Zah89] and Campbell and Papaloizou [CaP83] (see also Savonije and Papaloizou [SaP85] and Mazeh et al. [MLM90]), is based on Darwin's tidal torque theory and tidal friction produced by turbulent viscosity.

Recently a purely hydrodynamical mechanism involving transient meridional currents was proposed by Tassoul [Tas88]. These currents based on the Ekman advection of angular momentum should be much more efficient for the achievement of synchronization in early and late-type stars than any other known mechanism. We will come back to this theoretical aspect at the end of this paragraph and in the conclusions.

Computations based on theoretical developments should always be compared when possible with data derived from observations. The available data are:
1. The variation with time of the rotational velocity of single stars or binary components without orbit-spin coupling.
2. The frequency distribution of the periods for late-type MS binaries.
3. The circularization or, preferably, the degree of synchronization of AMLOSC binaries with well-known period.
The first point was already discussed in the second paragraph. Most observers find an exponential relation with a constant or slightly variable exponent ($c_4 \simeq 3$), as it was also discussed theoretically ([Mes84], [MeS87], [Ste88]) for stars with a rotational period $P \geq 2$. For young and fast rotators it is possible to derive (see [Sta91], also for references) a spin-down equation of $\log -\dot{\omega} = -19$ for GV MS stars. This result is in agreement with the spin-down relation derived from the frequency distribution of close binaries when we take a coupling coefficient $x \geq 0.5$.

The synchronization parameter mentioned in the third point also has a great diagnostic value. Figure 4 shows the effect of the different mechanisms on the degree of synchronization f. Moreover, we know from the tidal interaction theories that synchronization is achieved before circularization so that we can also use determinations of the orbital eccentricity for the study of synchronization (see however our conclusions). Spectroscopic observations of the eccentricity are often more easy to obtain for MS late-type stars and different groups of stars have been studied to determine the limiting period for which the eccentricity becomes definitely different from zero (see for ex. Mathieu et al. [MWM89] for the pre-main sequence stars, Jasniewicz and Mayor [JaM88], Latham [Lat92] and Duquennoy [Duq92] for the halo stars and Mayor and Mermilliod [MMe84] for coeval binaries of galactic clusters).

Moreover, f can be determined from a comparison of the orbital revolution and the rotation of the component(s) obtained by means of photometric observations of eclipsing binaries. All well-studied close late-type binaries show signs of magnetic activity and the dark spots connected with this activity generate a detectable modulation on the light curve of the eclipsing systems. From this modulation it is possible to measure the motion of the spots (supposed to rotate with the same velocity as the local photosphere) with respect to the orbital motion, and hence the degree of synchronization. From the results published (see for example [SHB89], also for references) or communicated to us until now we get the clear impression that the synchronization is much better than predicted by the classical tidal interaction theory.

This result would be in favour of the mechanism developed by Tassoul [Tas88] which gives much better synchronization. However this theory has been severely criticized during this workshop by Rieutord [Rie92], who finds that the mechanism proposed by Tassoul cannot exist in the stars.

However nothing is known about the structure of the magnetic fields of the components, let alone about the interaction between these fields and the possibility of magnetic locking ([DCa79],[Srl81]) which may also be strongly influenced by the geometry of the magnetic field. So, it still remains to be shown which mechanism may be at the origin of the good synchronization.

Concerning the magnetic field we must also say that its geometry will have a great influence on the AML by virtue of its action on the stellar winds. The change of the geometry with rotation has been studied by Roxburgh [Rox87], Vilhu and Moss [ViM86], Mestel and Spruit [MeS87], Taam and spruit [TaS89] and others. When the stellar rotation is rapid the magnetic field generated by the dynamo will adopt a more complex geometry and the magnetic braking effect will decrease. In reality we observe an increase of the spin-down which is about one order of magnitude higher than the increase expected from an extrapolation of the Skumanich relation. We think that a rapid increase of the mass loss, as it is also observed for the rapidly rotating young T Tauri stars, can explain this effect. Both types of stars will develop strong centrifugal winds at the surface, these winds are responsible for the AML [Mes84].

The switch from thermal winds to centrifugal winds at a period of 1 or 2d could be at the origin of the discontinuity found by us in the spin-down relation shown in Fig.1 ([Ste88]). In addition to the geometry of the magnetic field, which is difficult to model, the interaction between magnetic field and convection is strongly non linear.

4.   Conclusions

AMLOSC binaries are systems with AM losing components which also exercise a mutual tidal (magnetic) interaction such that transfer of AM from the orbit to the components and vice versa will take place (Orbit-Spin Coupling). From a systematic point of view there are still 3 other classes of binaries: nAMLOSC, AMLnOSC, nAMLnOSC, where n stands for "non".

Current tidal theories were developed for nAMLOSC binaries and it is clear that the synchronization problem is much less complicated (but not yet solved) for this group. The only action on the rotation of their components is produced by the companion. There are no magnetic torques acting on ionized stellar winds.

AMLnOSC binaries have components that are spinning down like single stars and no orbital change is expected (at least for the reasons discussed in this paper). They only pose us, as single stars do, the difficult problem to know how they spin-down! The nAMLnOSC binaries possess AM but keep it safe until the later stages of evolution.

In the present talk we were adding the problems of AML and OSC and we tried to learn about the one with the help of the other. The results are far from final:
1. A preliminary spin-down relation for the whole rotational velocity range.
2. A reasonable period evolution function (PEF) that gives the possibility to explain the presently observed period distribution and to compute the evolution from detached AMLOSC binary to single star via the contact binary stage.
3. The strong feeling that the interaction between the components is not yet completely understood.
4. Several reasons to believe that the convective zone of G-type MS stars is reasonably well coupled with the inner layers.

One of the important points that has become clear from the study of the AMLOSC binary is that the problem of synchronization can not be formulated as a simple (but already complicated) question of tidal interaction between components. Apart the unsolved question of magnetic interaction, the AML from the components can not be neglected.
The equations for the circularization and the synchronization time of late-type stars are based on the mechanism of turbulent friction in the convective zone. However this same convective zone is also the reason for the existence of AML, so it is not justified for this type of binaries to predict a synchronization time based on the turbulent friction alone. What happens in reality for AMLOSC binaries (see Fig.4) is that two synchronization events take place. The first, after about 10 My, is the result of spinning down of the companion caused by the combined effect of AML+AMT. This synchronization is transient for the AML will continue until equilibrium between AML and AMT is reached (AMT is now acting in the opposite sense). This is also the beginning of the approach of the two components and a gradual and definite increase of the synchronization parameter f.
This image is completely different from the nAMLOSC case for which the synchronization is gradually established by the tidal torque acting alone on the components which in the initial phase have a higher rotation than the revolution of the orbit and will never have a smaller rotation, at least according to the tidal interaction.

Acknowledgements

We are grateful to Michel Rieutord, Evry Schatzman, Jean-Louis Tassoul, Alexander Tutukov and Jean-Paul Zahn for suggestions and lively discussions. We also thank the italian and french Institutions which provided financial support to our collaboration: the Consiglio Nazionale delle Ricerche (CNR), the Observatory of Rome and the

Centre National de la Recherche Scientifique.

References

[CaP83] Campbell C.G., Papaloizou J., *The possibility of non-synchronism of convective secondaries in close binary stars.* Mon. Notices Roy. Astron. Soc., Vol. 204 (1983), pp. 433-447.

[DCB79] DeCampli W.M., Baliunas S.L., *What tides and flares do to RS Canum Venaticorum binaries.* Astrophys. J., Vol. 230 (1979), pp. 815-821.

[Duq92] Duquennoy A. *This workshop.*

[Egg82] Eggleton P.P., *Evolution of RS CVn and W UMa systems. Cool Stars, Stellar Systems and the Sun: 2nd Cambridge Workshop,* eds. M.S. Giampapa, L. Golub, (1982), pp. 153-159.

[FLM79] Farinella P., Luzny F., Mantegazza L., Paolicchi P., *Statistics of period for eclipsing binary systems.* Astrophys. J., Vol. 234 (1979), pp. 973-977.

[Hua66] Huang S.S., *A theory of the origin and evolution of contact binaries.* Ann. d'Astrophys., Vol. 29 (1966), pp. 331-338.

[JaM88] Jasniewicz G., Mayor M., *Radial velocity measurements of a sample of northern metal-deficient stars.* Astron. Astrophys., Vol. 203 (1988), pp. 329-340.

[Lat92] Latham D. *This workshop.*

[MaV91] Maceroni C., van 't Veer F., *The evolution and synchronization of angular momentum losing G-type main sequence binaries.* Astron. Astrophys., Vol. 246 (1990), pp. 91-98.

[Mes84] Mestel L., *Angular momentum loss during pre-main sequence contraction.* Proc. 3rd Cambridge Workshop on Cool Stars, Stellar Systems and the Sun, eds. S. Balunias, L. Hartmann, (1984), pp. 49-59.

[MeS87] Mestel L., Spruit H.C., *On magnetic braking of late-type stars.* Mon. Notices. Roy. Astron. Soc., Vol. 226 (1987), pp. 57-66.

[MLM90] Mazeh T., Latham D.W., Mathieu R.D., Carney B.W., *On the orbital circularization of close binaries. NATO-ASI on Active Close Binaries,* eds. C. Ibanoglu I. Yavuz, (1990), pp. 145-154.

[MMe84] Mayor M., Mermilliod J.-C., *Orbit circularization time in binary stellar systems. Observational tests of stellar systems evolution theory, IAU Symp. 105,* eds. A. Maeder, A. Renzini, (1984), pp. 411-414.

[MWM89] Mathieu R.D., Walter F.M., Myers P.C., *The discovery of six pre-main-sequence spectroscopic binaries.* Astron. J., Vol. 98 (1989), pp. 987-1001.

[OkS70] Okamoto I., Saito K., *The formation of W Ursae Majoris stars.* Publ. Astron. Soc. Japan, Vol. 22 (1970), pp. 317-333.

[Rah81] Rahunen T., *Evolution of W UMa systems and angular momentum loss.* Astron. Astrophys., Vol. 102 (1981), pp. 81-90.

[Rie92] Rieutord M., *This workshop.*

[Rox87] Roxburgh I., *Problems of the solar interior.* The Internal Solar Angular Velocity, eds. B.R. Durney, S. Sofia, (1987), pp. 1-5.

[SaP85] Savonije G.J., Papaloizou J.C.B., *Tidal interaction of close binary systems.* Interacting Binaries, eds. P.P. Eggleton, J.E. Pringle, (1985), pp. 83-102.

[Sch62] Schatzman E., *A theory of the role of magnetic activity during star formation.* Ann. d'Astrophys., Vol. 25 (1962), pp. 18-29.

[SHB89] Strassmeier K.G., Hall D.S., Boyd L.J., Genet R.M., *Photometric variability in chromospherically active stars. III. The binary stars.* Astrophys. J. Suppl., Vol. 69 (1989), pp. 141-215.

[Sku72] Skumanich A., *Time scales for CaII emission decay, rotational braking, and lithium depletion.* Astrophys. J., Vol. 171 (1972), pp. 565-567.

[Smi79] Smith M.A., *Rotational studies of lower main sequence stars.* Publ. Astron. Soc. Pacific, Vol. 91 (1979), pp. 737-745.

[Sod83] Soderblom D.R., *Rotational studies of late-type stars. II. Ages of solar type stars and the rotational history of the sun.* Astrophys. J. Suppl., Vol. 53 (1983), pp. 1-15.

[Srl81] Scharlemann E.T., *Tides in differentially rotating convective envelopes.* Astrophys. J., Vol. 246 (1981), pp. 292-305.

[Sta91] Stauffer J.R., *Rotational velocities of low mass stars in young clusters.* Angular momentum evolution of young stars, eds. S. Catalano and J.R. Stauffer, (1991), pp. 117-134.

[Ste88] Stepien K., *Spin-down of cool stars during their main-sequence life.* Astrophys. J., Vol. 335 (1988), pp. 907-913.

[TaS89] Taam R.E., Spruit H.C., *The disrupted magnetic braking hypothesis.* Astrophys. J., Vol. 345 (1989), pp. 972-977.

[Tas88]   Tassoul J.L., *On orbital circularization in detached close binaries*. Astrophys. J. Lett., Vol. 324 (1988), pp. L71-L73.

[VtV79]   van 't Veer F., *The angular momentum controlled evolution of solar type contact binaries*. Astron. Astrophys., Vol. 80 (1979), pp. 287-295.

[VeM88]   van 't Veer F., Maceroni C., *The angular momentum loss of rapidly rotating late-type main sequence binaries*. Astron. Astrophys., Vol. 199 (1988), pp. 183-190.

[VeM89]   van 't Veer F., Maceroni C., *The angular momentum loss for late-type stars*. Astron. Astrophys., Vol. 220 (1989), pp. 128-134.

[Vil82]   Vilhu O., *Detached⟶Contact scenario for the origin of W UMa stars*. Astron. Astrophys., Vol. 109 (1982), pp. 17-22.

[ViM86]   Vilhu O., Moss D., *Magnetic braking in cool dwarfs*. Astron. J., Vol. 92, (1986), pp. 1178-1182.

[Zah77]   Zahn J.-P., *Tidal friction in close binaries*. Astron. Astrophys., Vol. 57 (1977), pp.383-394.

[Zah89]   Zahn J.-P., *Tidal Evolution of close binary systems. I. Revisiting the theory of equilibrium tide*. Astron. Astrophys., Vol. 220 (1989), pp. 112-116.

Present State of the Tidal Theory

Jean-Paul Zahn *

To the memory of Arlette Rocca

1. Introduction

The observations of binary stars are progressing on a very fast pace, both in quantity and in quality. This is amply demonstrated again by the present meeting, which is held in honor of one of the prominent explorers of that exciting field. And, as often in astrophysics, such observational advances challenge the theoreticians to improve their share of the work.

Angular momentum is a (vector) quantity which is conserved in a closed system. (We shall ignore here the momentum which escapes through gravitational radiation, for it is in general extremely small, except in compact binaries.) Thus binaries can only lose angular momentum through a stellar wind, and that process can be significant indeed during certain phases of their evolution (see for instance the recent review by Hartmann and Noyes [HaN87]. Beside this possibility, there may be exchanges of angular momentum within the binary system, between the rotation of the stars and the orbital motion. These too may be accomplished by a transfer of matter from one component to the other, as it occurs in the later stages of close binary evolution. But such exchanges can also be mediated through gravitational interaction. Since they involve a change of the kinetic energy of the system, that energy must necessarily diminish, and therefore be dissipated into heat, as required by the second principle of thermodynamics.

Eventually, the binary may settle in its state of minimum kinetic energy, in which the orbit is circular, the rotation of both stars is synchronized with the orbital motion, and their spin axis are perpendicular to the orbital plane. Whether the system actually reaches this state is determined by the strength of the tidal interaction, thus by the separation of the two components, or equivalently the orbital period. But it also depends on the efficiency of the physical processes which are responsible for the dissipation of the kinetic energy.

---

*Observatoire Midi-Pyrénées, Toulouse, France, and Astronomy Department, Columbia University, New York, USA

Provided that we understand well enough these dissipation processes, the present state of a binary system can give us some precious clues on its past evolution, and possibly even on the conditions of its formation. This is one of the most interesting aspects of the problem, on which I wish to put some emphasis while reviewing what we know today about the tidal theory.

2.   The mechanism of tidal braking

The mechanism of tidal interaction has been well understood since Newton's time. The theory has of course been applied mainly to the Earth-Moon system, but later also to binary stars (Darwin [Dar79]). Detailed descriptions are given in many text-books, and in specialized treatises such as Kopal [Kop59]. Thus for the present purpose, a simple and mostly intuitive sketch will suffice.

2.1.   The basic mechanism

When the gravitational field of the companion star (of mass m) is superposed on the gravity of the primary star (of mass M and radius R), it produces there a tidal bulge whose relative thickness $\delta R/R$ is approximately the ratio of the (differential) perturbing force to the gravity at the surface. Thus we have

$$\frac{\delta R}{R} \approx \frac{GmR/a^3}{GM/R^2} = \frac{m}{M}\left(\frac{R}{a}\right)^3 , \tag{1}$$

a being the distance between the centers of the two stars. If the primary star were of constant density, its tidal bulge would have a mass of order $\delta M \approx (\delta R/R)M$; the actual figure is less, due to the density stratification. That tidal bulge produces a dipolar gravity field, which causes the motion of the apsides in an elliptic orbit [Kop59].

When the rotation of the star is synchronized with the orbital motion, the tidal bulge is perfectly aligned with the companion star. However, when there is no such synchronization, any type of dissipation causes a slight lag of the tidal bulge, and the star then experiences a torque which tends to pull it back to synchronism (see fig. 1). That torque Γ can be easily estimated:

$$\Gamma \approx -R\,\delta M \left(\frac{GmR}{a^3}\right)\sin\alpha \approx -\frac{Gm^2}{R}\left(\frac{R}{a}\right)^6 \sin\alpha, \tag{2}$$

where α is the tidal lag angle.

In the simplest case, that angle is proportional to the lack of synchronism $(\Omega - \omega)$, Ω being the rotation rate and ω the orbital angular velocity. It is also proportional to the strength of the physical process which is responsible for the dissipation of the

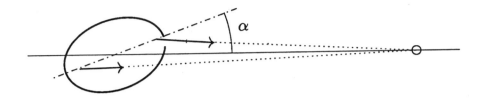

Figure 1. The torque exerted on the tidal bulge of a binary component. The companion is approximated by a point mass; all other features are grossly exagerated.

kinetic energy into heat, and therefore it is inversely proportional to the characteristic time t_f of that process. We are thus led to write

$$\alpha \approx \frac{(\Omega - \omega)}{t_f} \frac{R^3}{GM}, \tag{3}$$

where the most "natural" time, namely the free-fall time $(GM/R^3)^{-1/2}$, has been introduced to render α non-dimensional. Thus the synchronization rate is expected to be

$$\frac{1}{t_{sync}} = -\frac{1}{(\Omega - \omega)} \frac{d\Omega}{dt} = -\frac{\Gamma}{I(\Omega - \omega)} \approx \frac{1}{t_f} q^2 \frac{MR^2}{I} \left(\frac{R}{a}\right)^6, \tag{4}$$

where $q = m/M$ is the mass ratio and I the moment of inertia of the considered star.

The correct expression, which one obtains by solving the full equations governing the problem, is precisely of that form:

$$\frac{1}{t_{sync}} = 6 \frac{k_2}{t_f} q^2 \frac{MR^2}{I} \left(\frac{R}{a}\right)^6. \tag{5}$$

The apsidal constant k_2 measures the response to the external dipole field imposed by the companion; it is a function of the mass concentration within the star. Using Kepler's third law, one sees that the synchronization time varies as the 4th power of the orbital period.

Likewise, the time characterizing the circularization of the orbit is given by

$$\frac{1}{t_{circ}} = -\frac{d\ln e}{dt} = \frac{21}{2} \frac{k_2}{t_f} q(1+q) \left(\frac{R}{a}\right)^8; \tag{6}$$

the companion star contributes for a similar amount. (From now on, a designates the semi-major axis; e is the eccentricity.) The large difference between these two timescales reflects the disparity of the moments of inertia: that of the orbit is about Ma^2, whereas for the stars $I < MR^2$. It is for this reason that the synchronization of the components proceeds (in general) much faster than the circularization of the orbit.

2.2.   Equilibrium tide and dynamical tide

In deriving the expressions (5) and (6) above, one makes the implicit assumption that the binary component adjusts immediately to the potential exerted by the companion star. By using this hydrostatic approximation, one thus ignores the inertial effects in the response of the restoring forces to the tidal perturbation.

Among these forces, two come effectively into play in the tidal interaction, namely the buoyancy (in the stably stratified regions) and the Coriolis force. Both give rise to waves, which are reflected at the surface and can thus form standing waves. Only the buoyancy remains in the absence of rotation, and it is responsible for the so-called *gravity modes*, or inertial modes. Their eigenperiods range from about one hour to infinity. In a rotating star, these modes are somewhat altered by the Coriolis force; for instance, their frequencies are split in a way which is similar to the Zeeman effect in atomic physics (Cowling & Newing [CN49], Ledoux [Led51]).

If the rotating star were neutrally stratified, the Coriolis force would allow for another family of modes, the internal modes. However, only a fraction of them survives as such in a stably stratified star: these are the *quasi-toroidal modes* (Zahn [Z66b], Provost et al. [PBR79]. The motions associated with them are practically horizontal, and their eigenperiods are of the order of the rotation period.

Due to these restoring forces, the star behaves as an oscillator, and its reaction to a perturbing force does not reduce to an instantaneous, hydrostatic adjustment. Resonances may even arise, when the tidal frequency coincides with the eigenfrequency of one of the oscillation modes.

Thus the response of a star to the perturbing potential imposed by a companion star is made of two parts: the *equilibrium tide* and the *dynamical tide*. The latter is usually rather weak, and it does not contribute much to the tidal bulge, as was recognized already by Cowling [C41]. However, the oscillation associated with it involves radial scales that are small compared to the size of the star, and that can therefore enhance the dissipative processes, as we shall see later on.

3. The physical causes of tidal friction

Since the pionneering work of Darwin [Dar79], the mechanism of tidal braking has been extensively discussed in the astrophysical literature (see for instance Jeans [Jea29] and Kopal [Kop59]). It was the lack of knowledge concerning the physical causes of tidal friction which prevented from applying it to a more quantitative interpretation of observed binary systems. These causes were only identified starting in the mid-sixties, and work is still in progress to improve their description.

3.1. Turbulent dissipation

The viscosity of the stellar plasma is rather low: typically $\nu \approx 10 - 10^3$ cm^{2}s^{-1}. Therefore the timescale which characterises the viscous adjustment of a star as a whole, $R^2/\nu \approx 10^{12} - 10^{13}$ years, greatly exceeds the age of the Universe!

Nevertheless, viscous friction can play a major role in those regions of the star that are turbulent. And turbulence is very likely to occur in such a nearly inviscid medium. As is well known, whenever turbulence arises in a fluid, the kinetic energy of large scale flows cascades down to smaller and smaller scales until it can be dissipated into heat. It is a s if the viscosity had been greatly enhanced; one finds that the "turbulent viscosity" is of order $\nu_t \approx u\ell$, where u and ℓ are the typical velocity and size (or mean free path) of the turbulent eddies.

The major cause of turbulence in stars is thermal convection: the convective regions are highly turbulent, with Reynolds numbers reaching 10^{12}. Their turbulent viscosity can be easily estimated with the usual mixing-length treatment. For a star possessing a deep convective envelope, such as the Sun, in which most of the heat flux is transported by convection, one finds that the dissipation time-scale $t_{conv} = R^2/\nu_t$ is of the order [Z66a]

$$t_f = t_{conv} \approx \left(\frac{MR^2}{L} \right)^{1/3} , \qquad (7)$$

where L is the luminosity of the star. This time is rather short, about 1 year for the Sun.

In stars with a convective core, that friction time is strongly increased, because it scales as $(r_c/R)^{-7}$ with the size r_c of the core. Moreover, the turbulent viscosity can no longer be estimated in a straightforward way, because the turnover time of the convective eddies becomes longer than the tidal period (we shall discuss this point later in more detail, in §4.1). For both reasons, turbulent dissipation plays only a minor role in binaries whos e components do not possess a convective envelope.

It thus appears that we have an interpretation for the well established fact that close binaries with late-type components are obse rved with circular orbits. But how can we explain the behavior of early-type systems?

3.2.   Radiative damping

The other dissipation mechanism which comes into mind is radiative damping, and indeed it proves the most efficient process in the reg ions of the star which are not turbulent. Its characteristic time scale, for the star as a whole, is of the order of the Kelvin-Helmholtz time:

$$t_f = t_{rad} \approx \frac{GM^2}{RL} . \tag{8}$$

That global thermal relaxation time is much shorter than the viscous friction ti me, but not enough to account for the observed properties of early-type binaries.

In fact, radiative damping operates in a more subtle way. When an early-type star is not synchronized with the orbital motion, forced gravity waves are emitted from the lagging convective core. They carry kinetic energy towards the surface, where they are damped because the local thermal dissipation time become s there shorter than the tidal period.

For this reason, it is the *dynamical tide* which is the most efficient in these stars. To evaluate the relevant time scales, one has to calculate the amplitude of this tide. The problem to solve is that of the non-adiabatic gravity modes of the star forced by an outer potential, which is quite cumbersome when tackled head-on, as experienced by Savonije and Papaloizou [SaP83]. Fortunately, one can also apply asymptotic methods, because the excited modes are of high radial order [Z75a].

One finds that the torque does not vary linearly anymore with the tidal frequency (fig.2). It can become quite large at resonance with a low order gravity mode, but when the tidal period increases, as the star approaches synchronism, such resonances are smoothed out through radiative damping, because the outgoing gravity waves are increasingly absorbed. In contrast with the weakly damped equilibrium tide, the dynamical tide has turned into a purely propagating wave. The torque then scales as the power 8/3 of the tidal frequency $2(\Omega - \omega)$. That fractional exponent arises from the behavior of the gravity modes near the convective core, which is model dependent: in the original treatment the density gradient was assumed to be continuous. When there is a jump in that frequency just outside the convective core, such as due to convective penetration or to a discontinuous profile of the chemical composition, the torque scales as the cube of tidal frequency, as was shown by Rocca [Roc89].

It is thus appropriate to redefine the synchronization time as

$$\frac{1}{t_{sync}} = -\frac{d}{dt} \left| \frac{2(\Omega - \omega)}{\omega} \right|^{-5/3} = 5 \left(\frac{GM}{R^3} \right)^{1/2} q^2 (1+q)^{5/6} \frac{MR^2}{I} E_2 \left(\frac{R}{a} \right)^{17/2} , \tag{9}$$

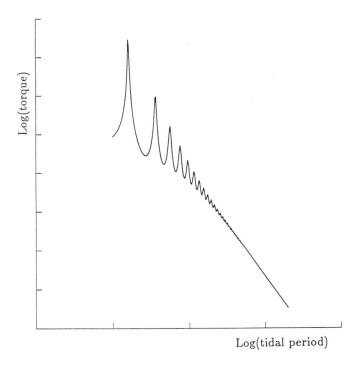

Figure 2. The tidal torque of the dynamical tide, in a star with a convective core and a radiative envelope. Note the resonances with the free modes of oscillation, which are damped out as the tidal period increases (from [Z75a]).

where we have used Kepler's third law to express ω in terms of the semi-major axis a. Comparing with eq. 5, we note that this expression does not involve any dissipation time scale. The structure constant E_2, which is the analog of the apsidal constant k_2, depends mainly on the size of the convective core; it is of order 10^{-6} for a 10 $M_\odot$ main-sequence star.

One obtains a similar expression for the circularization time scale:

$$\frac{1}{t_{circ}} = -\frac{d\ln e}{dt} = \frac{21}{2}\left(\frac{GM}{R^3}\right)^{1/2} q(1+q)^{11/6} E_2 \left(\frac{R}{a}\right)^{21/2} , \qquad (10)$$

which has to be completed with the contribution of the second star. We thus see that these timescales increase much more rapidly with the separation or the period than in the case of the equilibrium tide.

The main conclusion we can draw at this point is that there is a drastic difference in the strength of the tidal dissipation between stars which possess a convective envelope and those which do not. Let us illustrate this by an example. Imagine a (hypothetical) binary with a primary of 10 $M_\odot$ and a secondary of 1 $M_\odot$; both are on the main-sequence, with a radius of respectively 4 $R_\odot$ and 1 $R_\odot$. If the orbital period is 3 days, the synchronization time of the primary amounts to $5\,10^8$ yrs, thus exceeding its lifespan, but that of the secondary is only about $5\,10^5$ yrs. The circularization proceeds on a timescale of $\approx 2\,10^7$ yrs, and it is due almost entirely to the late-type secondary: the contribution of the primary is less than 10^{-3}.

The observations actually demonstrate that late-type binaries undergo stronger tidal dissipation than systems with early-type stars. Eclipsing binaries, which because of selection effects are in general very close systems, are found with circular orbits when they contain a late-type component [Z66a]; one exception is α CrB, but it is readily explained by the untypically large period of 17.36 days. For such stars, the circularization mechanism is so efficient that no eccentric systems survive, quite unfortunately, in which one could measure the apsidal motion.

A word of caution about the timescales we have derived above. They should be considered as first estimates, because they actually depend on other parameters, which need to be defined. For instance, the circularization time t_{circ} is a function of the rotation rates of the stars; the expressions given above (eqs. 6 and 10) apply once synchronization is achieved. Also, the synchronization times of eqs. 5 and 9 are valid only when the orbit is already circularized. If one wishes to describe the dynamical evolution of a binary system, one has to actually integrate in time the full set of equations governing the rotation of the two components, and the changes of orbital period and eccentricity; moreo ver, these equations should be coupled with a stellar evolution code to take into account the structural modifications of the stars, which are reflected in the value of most parameters involved: R, I, k_2, t_f, E_2. The outcome can differ substantially from that projected by the timescales above (see [Z77]). For instance, take a binary which has been completely relaxed on the main-sequence: its components will be despun, and thus desynchronized, as they evolve towards the giant branch; however the sy stem will keep a circular orbit.

4.   Refining the theory to meet the observations

The confrontation of the theoretical predictions with the observations bears mostly on the orbital eccentricities, and on the degree of synchronism of the two components, when it is measurable. The favored approach is to consider a sample of binaries, gathered on some criterion (cluster membership, for instance, which implies that the stars have the same age), and to examine whether there is a critical period below which the orbits are circular, and another below which the stars are synchronized.

4.1.   Late-type binaries

Until recently, it was felt that the observations agreed reasonably well with the circularization time scale predicted by eq. 5, the friction time being the convective time given by eq. 7. Koch and Hrivnak [KoH81] actually integrated the set of governing equations, from a large sample of initial coniditions on the zero age main-sequence (ZAMS), and they confirmed that the theoretical critical period for circularization was about 6 days (for solar-type stars of the solar age). This was less than the observed period, which is around 8 days, but the difference between the observed and theoretical time scales, barely an order of magnitude, was quite tolerable, considering the crude treatment of convection.

On the other hand, the observed rotational velocities of late-type binaries seemed also to agree fairly well with the predicted timescales for synchronization (eq. 6), as was verified by Giuricin et al. [GMM84c].

The relative sensitivity of the critical circularization period to the age of the stars, $P_{circ} \propto t^{3/16}$ according to this theory, prompted Mathieu and Mazeh [MaM88] to use it as a clock to measure the ages of galactic clusters. As they discuss it again in this meeting, they find some indication that the critical period varies from about 6 days for the youngest clusters ($\approx$ 100 Myrs) to 10 - 11 days for the oldest (10 Gyrs).

In order to refine the interpretation of these very promising measurements, I went back to the evaluation of the convective friction time, in the light of the new developments of the convective theory. When calculating the tidal torque, one has to perform a summation over the whole convection zone, taking into account the variation with depth of the turbulent viscosity; the correct expression had been derived in the meanwhile by Scharlemann [Sch82], who dispensed with an unnecessary approximation I made in the original treatment.

On the suggestion of Mayor and Mermilliod [MM84], I also wished to calculate the amount of circularization which occurs during the pre–main-sequence phase: there, the tidal interaction is much stronger, because the stars are of larger size, and also because they possess deeper convective envelopes. For this purpose, I had to address a delicate question, which had been ignored in the earlier work.

When the turnover time of the convective eddies becomes of the same order as the tidal period, one can no longer use the simple prescription $\nu_t \approx u\ell$ for the turbulent viscosity. I had encountered this problem in the convective core of the early-type binaries, where I thought reasonable to replace the mean free path ℓ of the eddies by the distance they cross during, say, half a tidal period [Z66a]. The turbulent viscosity is then given by

$$\nu_t = \frac{1}{3}u\ell \min[1, u\Pi/2\ell],$$ (11)

with Π being the tidal period.

A similar situation arises in the solar convection zone, when one needs a prescription for the turbulent damping of the 5-minutes oscillations. When Goldreich and Keeley [GK77] dealt with that case, they chose rightly to neglect the contribution of all turbulent eddies whose turnover time exceeds the considered period; the turbulent viscosity then becomes

$$\nu_t = \frac{1}{3} u\ell \min[1, (u\Pi/2\pi\ell)^2].$$ (12)

This expression was applied later to the tidal problem by Campbell and Papaloizou [CaP83], although the situation is somewhat different there.

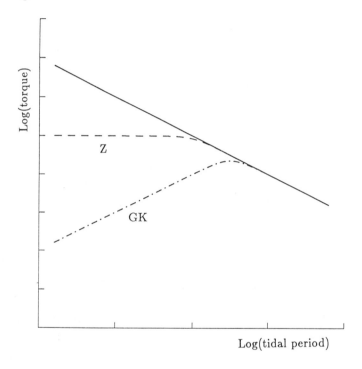

Figure 3. The tidal torque in a star with a convective envelope. The straight line corresponds to the unmodified turbulent viscosity, where the torque is proportional to the tidal frequency. When the tidal period becomes too short, that viscosity is reduced as explained in the text. The two other curves, labeled respectively Z and GK, show the result of implementing the prescriptions of eqs. 11 and 12 (From [Z89]).

These two prescriptions differ vastly for short tidal periods, as illustrated in fig. 3. The period at which the torque begins to depart from the linear regime increases as the convection zone deepens. For a 1 $M_\odot$ star, that transition period is about 8 days on the main-sequence, whereas it amounts to 40 days on the Hayashi track (with the prescription of eq. 11). For a system with solar-type components, the attenuation of the torque is thus rather slight on the main-sequence, but it could have been extremely important during the early phases of its evolution, depending on which prescription one believes.

Unfortunately, our knowledge about turbulent convection, moreover in a highly strat- ified medium, is still too limited to decide between the two prescriptions. The only indication comes from the Cepheid variables; as shown by Gonczi [Gon82], the obser- vations are better explained when implementing formula (11); the other, (12), yields a turbulent viscosity that is too weak to account for the red limit of the instability strip.

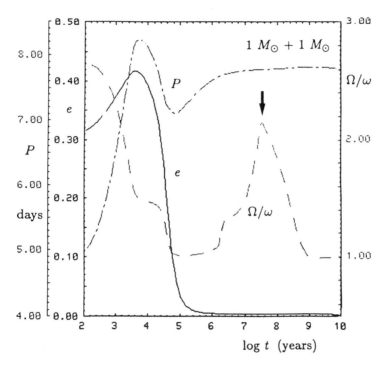

Figure 4. Synchronization and circularization of a binary system with two solar mass stars. The eccentricity e, the period P and the ratio Ω/ω between the rotational and orbital velocities are calculated using the prescription of eq. 11 for the turbulent viscosity. Note that most of the circularization occurs before the stars reach the zero age main-sequence, which is indicated by the arrow [ZaB89]; courtesy *Astron. Astrophys.*

With L. Bouchet, we integrated the full set of equations governing the tidal evolution, using the improved version of the friction time t_f [Z89] and prescription (11) for the reduction of the turbulent viscosity. To our surprise we found that, in spite of this reduction, most of the orbital circularization occurs in the pre–main-sequence phase, on the Hayashi track, when the stars are still entirely convective (Zahn & Bouchet

[ZaB89]). In other words, the eccentricity of late-type binaries is determined by the size and the separation they had when they were formed.

The result of one of our integrations is shown in fig. 4, for a binary with two components of 1 $M_\odot$. The initial separation has been chosen such that the eccentricity would decrease from 0.3, taken as initial value, to 0.005 on the ZAMS; the resulting binary has a period of $P_c = 7.8$ days. This is the critical circularization period of such systems: for larger periods, their eccentricity would exceed 0.005 on the ZAMS, and for smaller periods their eccentricity would not be detectable anymore. Note that the circularization is practically achieved at 10^6 yrs; the components are in synchronous rotation then, but they are spun up thereafter as they contract towards the ZAMS.

In this scheme, the initial conditions thus play a crucial role. Working back from the critical circularization period of about 8 days which is observed for field stars (Koch & Hrivnak [KoH81]), one can calculate the initial radii of the stars. They are in excellent agreement with those determined by Stahler [Sta83], [Sta88] through numerical simulations of the stellar formation, which account very well for the stellar birthline observed in the HR diagram.

The fact that early-type systems have a much smaller critical circularization period, less than 2 days according to Giuricin et al. [GMM84a], can also be explained. Palla and Stahler [PaS90] recently extented the stellar formation calculations to higher masses, and they found that stars with masses exceeding 2 $M_\odot$ are never fully convective in the pre–main-sequence phase, contrary to what was formerly thought. Therefore, binaries of that kind do not experience much tidal dissipation before they reach the main-sequence. This would mean that the contrasted behavior observed on the main-sequence, between binaries of lower and higher masses, is due more to their pre–main-sequence history than to the fact that late-type dwarfs possess a deep convective envelope, as advocated before.

If this scenario is correct, the critical circularization period of cluster members should not depend much on age. A weak increase is not ruled out, since the initial radii of stars change somewhat with the deuterium content of the protostellar material [Sta88]. It is therefore crucial to verify whether the trend detected by [MaM88] will be confirmed by more complete data (see this meeting, and also Goldman & Mazeh [GoM91]). If so, it would mean that the turbulent viscosity is two orders of magnitude less on the Hayashi track, which one could accommodate easily with the prescription (12) of Goldreich and Keeley. But one would also need to enhance the dissipation by a factor of order 100 on the main-sequence, and that appears more difficult to achieve, since the turbulent viscosity is closely linked to the convective flux, at least in the mixing-length treatment [Z89].

4.2.   Early-type systems

The eccentricities of early-type main-sequence binaries indicate that there is a critical period of about 2 days below which all orbits are circular, and above which both circular and elliptic orbits coexist (Giuricin et al. [GMM84b]). This is in reasonable agreement with the periods predicted by eq. 10, when replacing the circularization time by the lifespan of the stars on the main-sequence. Moreover, the fit may be improved by updating the value of the constant E_2 with the latest models, which have larger convective cores, and also by taking into account convective penetration (see Roxburgh [Rox89], [Z91]).

However, that circularization timescale was derived under two simplifying as sumptions, which are not fulfilled in the real star. First, the rotation had been neglected in calculating the dynamical tide. Subsequent work by Nicholson [Nic79] and Rocca [Roc87], [Roc89] included the Coriolis force, and correctly described the oscillation modes. They found that the vertical structure of the gravity modes is not much altered by the rotation, and therefore its impact on the dissipation timescale is rather limited.

Second, it was assumed that the star was behaving as a rigid body, whereas the tidal torque, which varies with depth, presumably induces some differential rotation. For this reason one expects that the surface layers are synchronized well before the deep interior [A83], [Z84]. The critical period derived from eq. 9, which is about 4 days, is based on a syn chronization timescale characterizing the whole star, and it does not take into account a possible decoupling of the outer layers from the deep interior. Actually, one observes that a large fraction of the early-type binaries appear to be synchronized beyond that critical period, as was pointed out by Rajamohan and Venkatakrishnan [RaV81], and by Giuricin et al. [GMM84a].

Only recently has this question been completely sorted out, thanks to Goldreich and Nicholson [GN89]. They showed that the rotation of such a binary component is synchronized gradually, layer after layer, starting from the surface, where radiative damping is the most important. The gravity waves emitted by the lagging convective core are absorbed at the co-rotation level, which deepens as the synchronization proceeds with time. Goldreich and Nicholson thus provide an illuminating explanation for the fact that many binaries appear to be synchronized well above the critical period inferred from eq. 9.

The apparent disagreement between these observations and the theoretical synchronization timescale incited Tassoul [Tas87] to propose another dissipation mechanism, taking place in the viscous boundary layer that develops in rotating fluids, namely the Ekman layer. It turns out, however, that this layer plays no effective role in the viscous dissipation when the boundary is of the stress-free type, which it is in the stellar case. This point is discussed by M. Rieutord during our meeting and a clarifying paper is being published [Rie92].

5.   Concluding remarks

The broad conclusion I would like to draw is that the tidal theory, in its present state, permits an interpretation of the observed properties of binary systems which is rather satisfactory. I take this for a sign that the main physical causes of tidal dissipation have been identified, since they operate on timescales which are compatible with those inferred from the observational data.

Does this rule out other mechanisms? Certainly not, because these may have escaped our attention so far if they intervene only at some stages of binary evolution. For example, there are some indications for weak turbulent motions in the radiation zone of rotating stars, due to the shear of horizontal differential rotation [Z75b], [Z87]. Such turbulence could well contribute to tidal dissipation, together with radiat ive damping, particularly in massive, very luminous stars.

But I feel that we should give priority to first work out the quantitative agreement of the present theory. For instance, we should implement the most recent models of early-type stars, and allow for convective penetration from the core. There is also a need for complete time integrations, coupled with a stellar evolution code, such as have been recently performed by Habets and Zwaan [HaZ89] or by [ZaB89]. Ideally, they should take into account the loss of angu lar momentum through stellar winds, and even accommodate differential rotation inside the stars.

Finally, let me insist again on the uncertainties of the convection theory, which is the base of the predictions concerning stars with convective envelopes. The latest numerical simulations reveal that thermal convection in a stratified medium is very intermittent, with strong plume-like downdrafts that carry most of the energy (Stein & Nordlund [StN89], Cattaneao et al. [CBT91]). We do not know how these plumes will interact with the tidal wave, and whether we can still use our concept of a turbulent viscosity, not to mentio n the difficulties which arise when the tidal period exceeds the characteristic turbulent time.

The observations of close binaries have been of paramount importance in the past to constrain the stellar structure theory. They provide the most accurate data on masses and radii. Their apsidal motion permits to validate models of stellar interior. Moreover, they offer sensitive tests for the evolution calculations, as was again demonstrated by J. Andersen during this meting (see also Andersen et al [ANC90]). But we may reap even more from the binary observations, if we succeed in refining our physical interpretation of the tidal interactions.

Acknowledgements. I wish to thank Ed Spiegel for kindly reading the manuscript, and for suggesting some improvements.

References

[ANC90] Andersen J., Nordström B., Clausen J.V. 1990, *Astrophys. J.* **363**, L33

[A83] Ando H. 1983, *Publ. Astron. Soc. Japan* **35**, 343

[CaP83] Campbell C.G., Papaloizou J.C.B. 1983, *Monthly Notices Roy. Astron. Soc.* **204**, 433

[CBT91] Cattaneo F., Brummel N.H., Toomre J., Malagoli A., Hurlburt N.E. 1991, *Astrophys. J.* **370**, 282

[C41] Cowling T.G. 1941, *Monthly Notices Roy. Astron. Soc.* **101**, 36

[CN49] Cowling T.G., Newing R.A. 1949, *Astrophys. J.* **109**, 149

[Dar79] Darwin G.H. 1879, *Phil. Trans. Roy. Soc.* **170**, 1

[GMM84a] Giuricin G., Mardirossian F., Mezzetti M. 1984a, *Astron. Astrophys.* **131**, 152

[GMM84b] Giuricin G., Mardirossian F., Mezzetti M. 1984b, *Astron. Astrophys.* **134**, 365

[GMM84c] Giuricin G., Mardirossian F., Mezzetti M. 1984c, *Astron. Astrophys.* **135**, 393

[GoM91] Goldman I., Mazeh T. 1991, *Astrophys. J.* **376**, 260

[GK77] Goldreich P., Keeley D.A. 1977, *Astrophys. J.* **211**, 934

[GN89] Goldreich P., Nicholson P.D. 1989, *Astrophys. J.* **342**, 1079

[Gon82] Gonczi G. 1982, *Astron. Astrophys.* **110**, 1

[HaZ89] Habets G.M.H.J., Zwaan C. 1989, *Astron. Astrophys.* **211**, 56

[HaN87] Hartmann L.W., Noyes R.W. 1987, *Ann. Rev. Astron. Astrophys.* **25**, 271

[Jea29] Jeans J. *1929, Astronomy and Cosmogony; Cambridge Univ. Press*

[KoH81] Koch R.H., Hrivnak B.J. 1981, *Astron. J.* **86**, 438

[Kop59] Kopal Z. *1959, Close Binary Systems; Chapman & Hall, London*

[Led51] Ledoux P. 1951, *Astrophys. J.* **114**, 373

[MaM88] Mathieu R.D., Mazeh T. 1988, *Astrophys. J.* **326**, 256

[MM84] Mayor M., Mermilliod J.-C. 1984, *Observational Tests of the Stellar Evolution Theory; eds. A. Maeder & A. Renzini, Reidel, Dordrecht, p. 411*

[Nic79] Nicholson P.D. 1979, Ph.D. thesis, Caltech

[PaS90] Palla F., Stahler S.W. 1990, *Astrophys. J.* **360**, L47

[PBR79] Provost J., Berthomieu G., Rocca A. 1979, *Astron. Astrophys.* **94**, 126

[RaV81] Rajamohan R., Vankatakrishnan P. 1981, *Bull. Astron. Soc. India* **9**, 309

[Rie92] Rieutord M. 1992, *Astron. Astrophys.* (to appear)

[Roc87] Rocca A. 1987, *Astron. Astrophys.* **175**, 81

[Roc89] Rocca A. 1989, *Astron. Astrophys.* **213**, 114

[Rox89] Roxburgh I.W. 1989, *Astron. Astrophys.* **211**, 361

[SaP83] Savonije G.J., Papaloizou J.C.B. 1983, *Monthly Notices Roy. Astron. Soc.* **203**, 581

[Sch82] Scharlemann E.T. 1982, *Astrophys. J.* **253**, 298

[Sta83] Stahler S.W. 1983, *Astrophys. J.* **274**, 822

[Sta88] Stahler S.W. 1988, *Astrophys. J.* **332**, 804

[StN89] Stein R., Nordlund Å. 1989, *Astrophys. J.* **342**, L95

[Tas87] Tassoul J.-L. 1987, *Astrophys. J.* **322**, 856

[Z66a] Zahn J.-P. 1966a, *Ann. Astrophys.* **29**, 489

[Z66b] Zahn J.-P. 1966b, *C. R. Acad. Sci.* **263**, 1077

[Z75a] Zahn J.-P. 1975a, *Astron. Astrophys.* **41**, 329

[Z75b] Zahn J.-P. 1975b, *Mém. Soc. Roy. Sci. Liège, 6e série, 31*

[Z77] Zahn J.-P. 1977, *Astron. Astrophys.* **57**, 383

[Z84] Zahn J.-P. 1984, *Observational Tests of the Stellar Evolution Theory; eds. A. Maeder & A. Renzini, Reidel, Dordrecht, p. 379*

[Z87] Zahn J.-P. 1987, *The Internal Solar Angular Velocity; eds. B. Durney & S. Sofia, Reidel, Dordrecht, p. 201*

[Z89] Zahn J.-P. 1989, *Astron. Astrophys.* **220**, 112

[Z91] Zahn J.-P. 1991, *Astron. Astrophys.* **252**, 179

[ZaB89] Zahn J.-P., Bouchet L. 1989, *Astron. Astrophys.* **223**, 112

Infrared Companions: Clues to Binary Star Formation

Hans Zinnecker *   *Bruce A. Wilking* †

Abstract

We discuss the nature of infrared companions found in close proximity to
pre-main-sequence (PMS) stars. We define an infrared companion as a member
of a PMS binary system which has a much larger infrared excess (e.g. K-L)
than the optically brighter star. We estimate that about 10% of all PMS binaries contain infrared companions, most of which are fairly wide pairs (100-300
AU). A large fraction of these systems are detectable in the 1.3mm continuum,
indicative of the presence of circumstellar dust. The discovery of several well-
known infrared companions is traced and a few new examples are described
(WSB4, Ser/G1, VV CrA). Either the binary systems with infrared companions are significantly younger than the average PMS star (10^5 yr instead of 10^6
yr) and have slightly non-coeval components (on the order of 10^5 yr), or such
systems have non co-planar circumstellar disks, misaligned with respect to the
orbital angular momentum. We propose future critical observations to help
discriminate between these possibilities.

1.   Definitions and Motivation

1.1.   Definitions

Infrared companions are companions to pre-main-sequence (PMS) or T Tauri stars
that are very red, to the extent that they are sometimes invisible in the optical.
Thus, the two components have significantly different spectral energy distributions,
and by implication different states of evolution (e.g., Adams, Lada, and Shu [ALS87]).
The infrared companion can sometimes be more luminous than the primary optical
component. But this does not necessarily mean that the infrared component is the
more massive component when the system arrives at the Main Sequence since much,
if not most, of the luminosity of the infrared companion could derive from accretion

*Würzburg University, W-8700 Würzburg, Germany
†University of Missouri-St. Louis, USA

of circumstellar matter onto the stellar surface. Given such different properties of the members of a physically bound binary system, questions are raised as to the coeval formation of the system, but, as we discuss below, a coeval origin of the components cannot be excluded at present.

1.2.   Motivation

The primary motivation to study these unusual systems comes from the suspicion that these might be *very* young binary systems so that we can hope to get closer to an understanding of binary formation. A second motivation is the possibility to learn about the dissipation of circumstellar matter (disks) around stars of the same or similar age but different stellar mass. Alternatively we may hope to prove that the circumstellar disks of the infrared companions are preferentially viewed edge-on. We can check if the statistics of the frequency of occurrence of infrared companions among all PMS binaries is consistent with random orientation of circumstellar disks.

2.   The History of Infrared Companions

The positions and projected separations for binary systems with infrared companions are presented in Table 1. Speckle interferometry from $\lambda = 1.25$ - 5 μm has played an important role in the discovery of infrared companions with projected separations from their primaries of 0.1″- 2.8″. The first IR companion to a T Tauri star was the companion to T Tau itself (Dyck et al. [DSZ82]). A third member of the system has been proposed, but its presence was not confirmed by recent 2-D speckle observations (Ghez *et al.* [GNG91]). It was not until 1986 that the second and third IR companions were found: Elias 22/GSS31 and Glass-I in the nearby ρ Ophiuchi and Chamaeleon dark clouds (Zinnecker et al. [ZCC87], Chelli et al. [CZC88]). Infrared companions were next identified with T Tauri stars in the Taurus cloud: Haro 6-10 (Leinert and Haas [LeH89]), XZ Tau (Haas, Leinert, and Zinnecker [HLZ90]), and UY Aur (Leinert, priv. communication). More recently, 2-D speckle observations have revealed an infrared companion to the FU Orionis class star, Z CMa, at a separation of 0.10 ″or 115 AU (Koresko et al. [KBG91], Christou et al. [CZH92]).

Furthermore, the high resolution imaging capabilities of infrared cameras have made it possible to study pre-main-sequence binaries with projected separations of 0.8″- 15″. For example, investigations of known binary T Tauri stars using direct infrared imaging (image scale 0.92″/pixel) have been made by Moneti and Zinnecker [MoZ91] and have led to the discovery that UZ Tau/e has a larger infrared (J-K) excess than UZ Tau/w. Using ProtoCAM at the Infrared Telescope Facility, a pixel scale of 0.135″can be chosen to resolve subarcsec binaries. New companions revealed by infrared cameras are also listed in Table 1 and include WSB4 in Ophiuchus (shown in Fig.1), Ser/G1 in Serpens, and VV Cr A in Corona Australis. Of these objects,

Table 1. Pre-Main-Sequence Stars with Infrared Companions. References for discovery of infrared companions and combined 1.3 mm flux densities: (a) [GNM92]; (b) [BSC90]; (c) [LeH89]; (d) [ReZ92]; (e) [HLZ90]; (f) [MoZ91]; (g) Leinert (priv. communication); (h) Ressler and Shure (priv. communication); (i) [KBG91]; (j) [WSD91]; (k) [CZC88]; (l) Zinnecker (unpublished data); (m) [MWZ92]; (n) André (priv. communication); (o) [WiZ92]; (p) [ReZ92].

Source	R.A. (1950)	Decl. (1950)	d (pc)	a (″)	a (AU)	S_ν(1.3 mm) (mJy)	Ref.
(1)	(2)	(3)	(4)	(5)	(6)	(7)	(8)
T Tau	$04^h19^m04.2^s$	$+19°25'05''$	140	0.66	90	280	(a)(b)
Haro 6-10	04 26 21.9	+24 26 30	140	1.2	170	116	(c)(d)
XZ Tau	04 28 46.0	+18 07 35	140	0.30	42	<18	(e)(b)
UZ Tau	04 29 39.3	+25 46 13	140	3.6	505	172	(f)(b)
UY Aur	04 48 35.7	+30 42 14	140	0.88	125	48	(g)(b)
SSV 63	05 43 34.4	−00 11 17	450	2.5	1125	847	(h)(d)
Z CMa	07 01 22.5	−11 28 36	1150	0.10	115	489	(i)(j)
Glass-I	11 06 50.2	−77 17 31	140	2.8	390	50	(k)(l)
WSB4	16 15 47.0	−26 02 54	160	2.7	430	<100	(m)(l)
Elias22/GSS31	16 23 22.0	−24 14 14	160	2.3	360	65	(k)(n)
Ser/G1	18 25 41.3	−00 04 39	500	2.9	1450	<100	(o)(l)
VV Cr A	18 59 43.9	−37 17 15	130	2.1	275	550	(p)(l)

UZ Tau and Ser/G1 had been previously identified as optical pairs (Herbig [Her62], Cohen and Kuhi [CoK79]).

3.   Possible Origins

There are two main working hypotheses which we can use to understand the nature of binary systems with infrared companions. The first set of hypotheses assumes the systems formed coevally to within 10^5 years. Then the strong infrared excess of only one member of the system would require either (1) slightly different masses and hence different evolutionary timescales for the dissipation of circumstellar dust or (2) a basic misalignment of their circumstellar disks assuming they are well-separated. The second hypothesis would be that the systems have similar evolutionary timescales but formed at slightly different times, with the youngest component still most heavily obscured.

The test of these hypotheses will come from the observed properties of these systems. Magnitude-limited unbiassed infrared surveys for binaries with infrared companions have been carried out and preliminary results are available (Zinnecker et al. [Zin92], Leinert et al. [LWH92], Ghez et al. [GNM92]). Although preliminary, these results

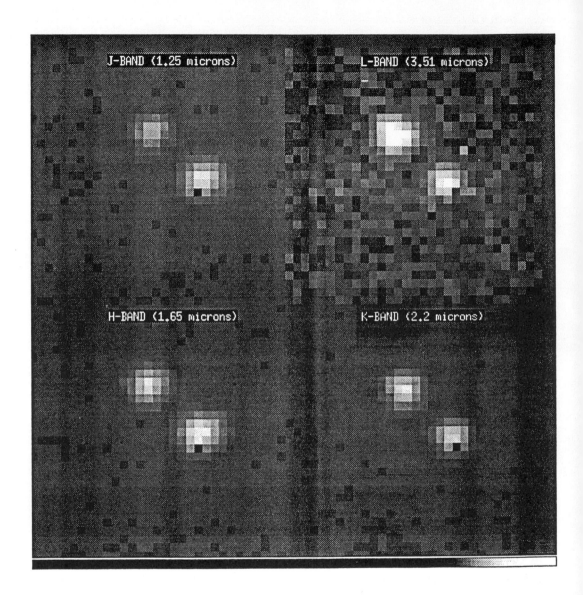

Figure 1. Infrared images of the WSB4 system at 1.25 and 1.6 μm (top) and 2.2 and 3.5 μm (bottom). These images were obtained with ProtoCAM, the imaging system at the Infrared Telescope Facility in Hawaii. The pixel scale is 0.35''/pixel. Note that the optically brighter star (northwest component) appears to have a smaller infrared excess and is fainter than the southeastern star at 2.2 and 3.5 μm [MWZ92].

can be used to infer some general ideas about systems with IR companions. First, their frequency of occurrence is small, comprising only ~10% of all T Tauri binary systems. If systems with infrared companions are precursors to binary T Tauri stars, then the "mean" age of T Tauri stars (e.g., 10^6 years) implies that these systems are extremely young, perhaps on the order of 10^5 years. It is interesting to point out that this is the estimated duration for the embedded Class I phase. Therefore, adopting the scenario proposed by Adams, Lada, and Shu [ALS87], a system with an infrared companion may be comprised of a primary which has dissipated its infalling envelope with a more slowly evolving companion which may still be in the accretion phase. If the system is coeval, then we must postulate that the primary is the more massive of the two objects and is the first to cross the "birthline" (e.g., Stahler [Sta83]). The shorter timescale observed for more massive stars to dissipate their circumstellar dust could be due to an earlier or stronger stellar wind, perhaps related to a mass-dependent onset of deuterium burning.

Consistent with this picture is that a somewhat higher fraction of systems with infrared companions display 1.3 mm continuum emission from cool circumstellar dust than do isolated T Tauri stars. In column 7 of Table 1 we list the observed flux densities at 1.3 mm for the combined systems. Note that 8 of the 11 systems (73%) have thus far been detected. For comparison, in their 1.3mm survey of 86 T Tauri stars, Beckwith *et al.* [BSC90] detected 53% of CTTS and 29% of WTTS. While this emission is unresolved at present, it could arise due to the presence of the embedded Class I companion or due to a circumbinary component associated with a young PMS system just emerging from the embedded state.

Infrared and optical observations suggest that extreme variability may be a characteristic of these systems. The primaries for both T Tau and VV Cr A have been observed to display brightness variations of several magnitudes in the optical and infrared, respectively (Ghez et al. [GNG91]; Reipurth and Wilking, private communication). Recently it has been observed that in the Z CMa system, the infrared excess can be highly variable on the timescale of one year (Christou, priv. communication) and may reflect a highly variable accretion rate. Recent infrared imaging of the Haro 6-10 system by Ressler and Shure (priv. communication) shows that the primary is not only variable but now 0.36 mag brighter than the infrared companion at 5 μm.

A test for the coevality might be to determine spectral types and luminosities for both members of a system where, despite one object having a large infrared excess, both members are visible (e.g., WSB4, Ser/G1). Then one could use a set of PMS isochrones to see if the objects are coeval. The possible pitfall is getting a proper luminosity for the photosphere of the object with the infrared excess, as a significant fraction of its luminosity could still be derived from accretion.

Finally, we note that the majority of systems in Table 1 have projected separations greater than 100 AU. This would seem to rule out the fragmentation of disks with diameters of 100-200 AU as the sole mechanism for forming infrared companions in this sample (Shu *et al.* [STA90]).

4. Future Observations

A number of critical observations can be carried out in the near future, in order to elucidate further the nature of the systems with IR companions. These observations will help to discriminate, for example, between the coeval and non-coeval hypotheses discussed above and will also address the issue of misaligned circumstellar disks. First, direct IR imaging at 3.1 μm and 9.7 μm should be able to pinpoint the strengths of the components' ice and silicate features. If, as expected, the optical depths of the ice and silicate absorption are larger for the IR companion than for the primary, we could infer a higher column density (mass) of circumstellar dust around the companion. Note that such differential measurements are insensitive to foreground absorption, since the lines of sight to the two components are almost the same. While the ice absorption measures cool dust (cooler than 125 K, the melting temperature of the ice mantles), silicate absorption is sensitive to dust at all temperatures up to the silicate grain evaporation temperature of about 1500-1700 K. The presence of silicate absorption has been suggested as a diagnostic to indicate Class I sources with infalling protostellar envelopes. It could also indicate Class II sources with nearly edge-on disks. The absence of silicate absorption does not imply that there is no circumstellar disk, it just implies that such a disk is not edge-on. Thus, it is conceivable that in systems with IR companions we will detect strong silicate absorption from the edge-on disk around IR companion but no absorption from the more face-on disk of the primary. Such misaligned circumstellar disks may explain the data, especially if both components had excess emission in the thermal infrared. For example, this model could apply to the T Tau system, where the IR companion T Tau S exhibits the silicate feature in absorption while on the optical star T Tau N it is in emission (Ghez et al. [GNG91]).

Broad-band 10 μm imaging would also be helpful for systems which have been detected by IRAS: it allows us to disentangle the IRAS 12 μm flux of the combined system, and by implication also the unresolved 25, 60, and 100 μm fluxes. This is an important pre-requisite for more correctly splitting up the combined luminosity of the system into the individual contributions of its components.

These thermal IR observations should be supplemented by optical and NIR polarimetry. The comparison of the degree, and especially the P.A., of scattered polarized light contains information about the alignment of the respective circumstellar disks (cf. Bastien and Ménard [BaM90]). Whether the disks within a system are coplanar could also be probed if we could observe jets and HH-objects from each component. This presupposes that jets always come out perpendicular to circumstellar disks and that both components have circumstellar disks. Recently, Reipurth et al. (unpubl.) may have observed two jet systems from a deeply embedded binary system (only seen at radio wavelengths), associated with the exciting source of the HH 1/2 jet system. The second jet system is not parallel to the HH 1/2 system, suggesting that the corresponding disks of the fiducial protobinary system are not aligned.

Future HST observations may reveal additional examples of young binaries with very small separations. They may also determine whether UV boundary layer emission comes from one or both components of an IR companion system, revealing if they are accreting and thus if their respective integrated luminosities are purely stellar or not. The question of accretion disks around both components is likely to be a subject for future interferometry with arrays of mm-telescopes (eg. MMA). For example, Simon and Guilloteau [SiG92] have recently resolved the dust continuum emission at 2.6mm in the GG Tau and UZ Tau binary systems. Interestingly, both systems are most likely hierarchical quadruple systems (Leinert et al. [LHR91], Martin et al. [MRM92]); i.e. their wide components are themselves binary systems. These hierarchical systems are a warning that the infrared excess of an infrared companion may in fact not result primarily from a circumstellar disk, but from yet another very cool unresolved companion (cf. Zinnecker [Zin89], p. 462). This unusual possibility would gain support if infrared absorption lines from cirumstellar dust cannot be detected toward infrared companions or if mm continuum emission is not seen by interferometric measurments. We are lucky that there are quite a few examples of IR companions with separations exceeding 1″, so that these new observations separating the IR companion from the primary component are indeed within reach.

Acknowledgements. We would like to thank Michael Meyer for providing us with Figure 1. We gratefully acknowledge support from a NATO collaborative research grant CRG 910174. One of us (HZ) also acknowledges funding from the DFG under Yo 5/7-1. Participation in the Bettmeralp meeting would have been impossible without the generous financial aid provided by the organizers. We are looking forward to a future similar meeting on the same topic and in the same location.

References

[ALS87] Adams, F. C., Lada, C. J., and Shu F. H. *1987, Ap. J. 312, 788*

[BaM90] Bastien, P. and Ménard, F. *1990, Ap. J. 364, 232*

[BSC90] Beckwith, S. V. W., Sargent, A. I., Chini, R. S., and Guesten, R. *1990, A. J., 99, 924*

[CoK79] Cohen, M. and Kuhi, L. V. *1979, Ap. J. Suppl. 41, 743*

[CZC88] Chelli, A., Zinnecker, H., Carrasco, L., Cruz-Gonzales, I., and Perrier, C. *1988, Astr. Ap. 207, 46*

[CZH92] Christou, J. C., Zinnecker, H., Haas, M., Leinert, C., and Ridgway, S. T. *1992, in High Spatial Resolution Imaging by Interferometry II, Ed J. Beckers and F. Merkle, (ESO Garching, in press)*

[DSZ82] Dyck, H. M., Simon, T., and Zuckerman, B. *1982, Ap. J. (Letters) 255, L103*

[GNG91] Ghez, A. M., Neugebauer, G., Gorham, P. W., Haniff, C. A., Kulkarni, S. R., and Matthews, K. *1991, A. J 102, 2066*

[GNM92] Ghez, A., Neugebauer, G., and Matthews, K. *1992, in Complementary Approaches to Double and Multiple Star Research, Ed W. Hartkopf and H. McAlister, (IAU Coll. 135, in press)*

[HLZ90] Haas, M., Leinert, Ch., and Zinnecker,H. *1990, Astr. Ap. 230, L1*

[Her62] Herbig, G.H. *1962, Adv. Astr. Ap. 1, 47*

[KBG91] Koresko, C. D., Beckwith, S. V. W., Ghez, A. M., Matthews, K., and Neugebauer, G. *1991, A. J. 102, 2073*

[LeH89] Leinert, Ch. and Haas, M. *1989, Ap. J. (Letters) 342, L39*

[LHR91] Leinert, Ch., Haas, M., Richichi, A., Zinnecker, H., and Mundt, R. *1991, Astr. Ap. 250, 407*

[LWH92] Leinert, Ch., Weitzel, N., Haas, M., Lenzen, R., Zinnecker, H., Christou, J., Ridgway, S., Jameson, R., and Richichi, A. *1992, in Complementary Approaches to Double and Multiple Star Research, Ed W. Hartkopf and H. McAlister, (IAU Coll. 135, in press)*

[MRM92] Martin, E. L., Rebolo, R., and Magazzu, A. *1992, in Complementary Approaches to Double and Multiple Star Research, Ed W. Hartkopf and H. McAlister, (IAU Coll. 135, in press)*

[MoZ91] Moneti, A., and Zinnecker, H. *1991, Astr. Ap. 242, 428*

[MWZ92] Meyer, M. R., Wilking, B. A., and Zinnecker, H. *1992, A. J., submitted*

[ReZ92] Reipurth, B. and Zinnecker, H. *1992, Astr. Ap., in preparartion*

[SAL87] Shu, F. H., Adams, F. C., and Lizano, S. *1987, Ann. Rev. Astr. Ap. 25, 23*

[STA90] Shu, F. H., Tremaine, S., Adams, F. C., and Ruden, S. P. *1990, Ap. J. 358, 495*

[SCH92] Simon, M., Chen, W. P., Howell, R. R., Benson, J. A., and Slowik, D. *1992, Ap. J. 384, 212*

[SiG92] Simon, M., and Guilloteau, S. *1992, Ap. J. Lett., submitted*

[Sta83] Stahler, S. W. *1983, Ap. J. 274, 822*

[WSD91] Weintraub, D. A., Sandell, G., and Duncan, W. D. *1991, Ap. J. 382, 270*

[WiZ92] Wilking, B. A. and Zinnecker, H. *1992, Ap. J., in preparation*

[ZCC87] Zinnecker, H., Chelli, A., Carrasco, L., Cruz-Gonzales, I., and Perrier, C. *1987, IAU Symposium 122, Ed I. Appenzeller and C. Jordan, 117*

[Zin89] Zinnecker, H. *1989, in Low-Mass Star Formation and Pre-Main Sequence Objects, Ed B. Reipurth, ESO Garching 447*

[Zin92] Zinnecker et al. *1992, in High Spatial Resolution Imaging by Interferometry II, Ed J. Beckers and F. Merkle, (ESO Garching, in press)*

THE DISTRIBUTION OF CUTOFF PERIODS WITH AGE: AN OBSERVATIONAL CONSTRAINT ON TIDAL CIRCULARIZATION THEORY

Robert D. Mathieu*   Antoine Duquennoy†

David W. Latham◊   Michel Mayor†

Tsevi Mazeh∇   Jean-Claude Mermilliod#

Abstract

Recent extensive radial-velocity surveys of several populations of solar-mass stars have provided orbital elements for large numbers of binaries in well-defined samples. These surveys are discussed in previous papers in this volume. This paper draws together those results and considers their implications for present theories of tidal circularization.

In particular, one property of these binary populations is that the shortest period orbits are typically circular. The range in period of these circular orbits can be delineated by a "circularization cutoff period", defined here as the period of the longest period circular orbit. Differing tidal circularization theories differ in their predictions of how these cutoff periods should evolve in time.

We first discuss critical uncertainties in comparison of the observed cutoff periods both with each other and theory. We then consider the observed dependence of cutoff periods with age of the binary samples. The cutoff periods are found to increase with time, indicative of effective main-sequence tidal circularization. The rate of increase is fully consistent with the predictions of classical tidal circularization theory. However, recently it has been suggested that at short periods the efficiency of turbulent dissipation depends on the binary period, leading to a stronger dependence of the cutoff period on age than in the classical theory. The observed cutoff periods are also consistent with this hypothesis, *if* the cutoff periods in the younger binary populations have been set by pre-main sequence tidal circularization. Such pre-main sequence tidal circularization is permitted but not required by present observations.

* Department of Astronomy, University of Wisconsin, Madison, Wisconsin 53706 USA

† Observatoire de Geneve, 51 chemin des Maillettes, CH 1290 Sauverny Switzerland

◊ Harvard-Smithsonian Center for Astrophysics, 60 Garden St., Cambridge, Massachusetts 02138 USA

∇ The Wise Observatory, Tel Aviv University, Ramat Aviv, Tel Aviv 69978 Israel

Institut d'Astronomie de l'Universite de Lausanne, CH 1290 Chavannes des Bois Switzerland

1. Introduction

It has long been known that the vast majority of binary stars have eccentric orbits, but that the shortest period binaries typically have circular orbits.[1] This result holds true across stellar mass and most evolutionary states. Usually the circularity of the short-period orbits has been attributed to stellar tidal circularization, as the efficiency of this process is greater for smaller stellar separations and thus shorter orbital periods. Indeed, while several theories of tidal circularization are under consideration at present, all agree that the timescale of circularization depends strongly on the separation of the two stars. Consequently, in a coeval sample of binaries the transition with increasing period from predominantly circular orbits to a distribution of eccentric orbits should occur at a characteristic or "cutoff" period.

Such transitions have been observed in the eccentricity distributions of several samples of approximately solar-mass binaries, on which we focus the discussion in this paper. Koch and Hrivnak [KoH81] examined the eccentricity distribution of F and G binaries in several compilations of binary orbits; with the exclusion of two anomalies which they noted, their distribution shows predominantly circular orbits for periods less than 10 days. Mayor and Mermilliod [MaM84] first studied the eccentricity distribution of coeval samples of binaries in open clusters. They also found a distinct cutoff period in the eccentricity distribution of Hyades/Praesepe binaries, but at a shorter period of 5.7^d. Mathieu, Latham and Griffin ([MLG90], cited earlier in [MaM88]) found a well-defined cutoff period at 10.3^d–11.0^d in the older open cluster M67. Similar transitions have been found among binaries in the field. Duquennoy and Mayor [DuM91] found a cutoff period of approximately 11.6^d among solar-mass binaries in the solar vicinity. Among galactic halo binaries, Jasniewicz and Mayor [JaM88] and Latham *et al*. [LMC88] found a transition between 12^d and 19^d. Finally, Mathieu, Walter and Myers [MWM89] have suggested a tentative cutoff period among pre-main sequence binaries near 4^d.

An important result of these studies was that longer cutoff periods were always found among older samples of binaries, as would be expected if tidal circularization processes were acting while the binary components were on the main sequence. Furthermore, Koch and Hrivnak [KoH81] and Mayor and Mermilliod [MaM84] argued that the cutoff periods found among field binaries and the Hyades/Praesepe binaries, respectively, were reasonably consistent with theoretical rates of tidal circularization of Zahn ([Zah66], [Zah77]) and Lecar, Wheeler and McKee [LWM76]. Mathieu and Mazeh [MaM88] noted that comparison of the Hyades/Praesepe and M67 cutoff periods suggested a weaker dependence of the rate of tidal circularization on period than predicted by Lecar, Wheeler and McKee [LWM76]. Zahn [Zah89] suggested a reduced efficiency of turbulent viscosity in short-period binaries, producing such a weaker dependence. Goldman and Mazeh [GoM91] argued that the reduction in viscosity was greater still, and claimed better agreement with the observed distribution of cutoff periods with age. Alternatively, Tassoul ([Tas87], [Tas88]) has argued for tidal circularization theory through large-scale, transient currents that also can produce agreement with the distribution of cutoff periods.

A fundamental change to the picture was suggested by Zahn and Bouchet [ZaB89], following an insightful comment of Mayor and Mermilliod [MaM84]. Based on the low rate of tidal circularization suggested by Zahn [Zah89], they argued that *all* tidal

---

[1] In this paper "circular orbits" is taken to mean orbits whose measured eccentricity is indistinguishable from zero, in contrast to "eccentric orbits".

circularization occurs during the pre-main sequence stage of evolution, when the stars have both larger radii and deeper convective zones, with no significant circularization occurring subsequently on the main sequence. Thus they predict that all samples of late spectral type binaries, regardless of age, should have similar cutoff periods between 7.2 and 8.5 days. On the other hand, Mazeh *et al.* [MLM90] recently reviewed the observations and concluded that in fact cutoff periods do evolve with age among main-sequence binary populations and that main-sequence tidal circularization is effective in main-sequence binaries.

Clearly, tidal circularization theory is presently in a state of flux, for which observations can provide guidance. The number of determined binary orbits is advancing rapidly. The primary goal of this paper is to summarize the present observational picture of the distribution of cutoff period with binary age, based on the results presented at this meeting. We also discuss the uncertainties in these measured cutoff periods. Finally, we compare the observed distribution of cutoff period with age against the several tidal circularization theories.

2. The Observed Distribution of Cutoff Period with Binary Age

In earlier papers in this field, the cutoff period was often discussed as if it were clearly delimited on the low side by the longest period circular orbit and on the high side by the shortest period eccentric orbit. However, this need not be the case even in a coeval sample of binaries due to differing initial eccentricities, primary and secondary masses, etc. ([MaM88], [DMM92]). For example, it is possible to find a binary with an eccentric orbit having a period shorter than another binary with a circular orbit, if the former binary had an *initial* eccentricity substantially larger than the latter. Mathieu and Mazeh [MaM88] estimated that a distribution in initial eccentricities would lead to a width in period of the transition region of order 14%, but they did not include period evolution during circularization. The significance of this effect has been shown by Duquennoy, Mayor and Mermilliod (hereinafter DMM; [DMM92]). Their calculations including both eccentricity and period evolution have found that orbits with small eccentricities might be found at periods half that of the longest period circular orbit.

Such widened transition regions have been observed. In the Hyades[2] an eccentric orbit is found at a period of 5.7^d while the longest period circular orbit has a period of 8.5^d [DMM92]. The longest period circular orbit in M67 has a period of 12.4^d, while an eccentric orbit is found at 11.0^d [LMG92]. However, it is notable that only in the Hyades are eccentric orbits found with periods substantially shorter than that of the longest period circular orbit, and in the Hyades the pedigree of the binary with the longest period circular orbit is not established (see below). In five other binary samples (Table 1), overlaps in period of circular and eccentric orbits are non-existent or small.

Nonetheless, given some cases of overlap a modified definition of the "cutoff period" is needed. Here we follow DMM in defining the cutoff period as the period of the longest period circular orbit. The presently known cutoff periods and ages are given in Table 1, summarizing the many studies presented at this meeting. Figure 1 (Section 4) presents this distribution graphically. In Table 1 we have followed Duquennoy and Mayor [DuM91] in

---

[2] For brevity, the Hyades/Praesepe sample will be referred to as "Hyades".

Table 1: Distribution of Observed Cutoff Periods with Age

Log Age (Gyr)	Cutoff Period	Binary Sample	Reference
-2.5	4.3^d	Pre-main sequence	[Mat92]
-1.0	7.05	Pleiades	[DMM92]
-0.1	8.5	Hyades/Praesepe	[DMM92]
0.6	12.4	M67	[LMG92]
<1>	10.3	Solar vicinity	[DuM91]
1.2	18.7	Halo	[LMT92]

adopting an old disk age of 7-11 Gyr for the solar-mass field binaries. Unfortunately, this sample includes a very large range of ages. The age of the binary defining the cutoff period (HD13974) has been given by Soderblom [Sod83] as 1.6×10^9 yr, but the reliability of the age has been questioned (see [DuM88]). Thus the age ambiguities in the sample are too large to make this cutoff period useful for our purposes here, and this datum is excluded from Figure 1. We have adopted 15 Gyr for the age of the halo sample. The age calibration of the galactic halo is not settled at the moment, with estimates in age differing by up to 5 Gyr. The ambiguity here is smaller than for the field solar-mass binaries and we have chosen to include the halo datum in our analysis; nonetheless this uncertainty in age should be kept in mind. Finally, the age adopted for the pre-main sequence sample is that of the binary with the longest period circular orbit (Lee and Mathieu, in preparation).

It is worth noting that the differences between the cutoff periods in Table 1 and those given in some previous publications for the same samples are in several cases the result of different definitions rather than new data. For example, Mazeh *et al.* [MLM90] define the cutoff period as that period such that all binaries with substantial orbital eccentricity (e.g., greater than 0.1) have longer periods (i.e. defining the cutoff period by the shortest period eccentric orbit). If the period distribution of short-period eccentric orbits is also set by tidal circularization processes, then that definition and the one used here are simply using different markers of the transition region from circular to eccentric orbits. If the parent eccentricity distributions in such transition regions do not differ substantially between binary samples, for large sample sizes the conclusions derived from either definition should be the same.

3. Uncertainties in the Cutoff Periods

Optimally the binaries from which the cutoff periods in Table 1 are derived would differ only in their ages and periods. Initial eccentricities, masses, stellar angular momenta, compositions, etc. would be identical, thus producing well-defined transition periods which would differ between samples only due to the influence of age. In addition, sample sizes would be sufficient to clearly delineate the cutoff periods. In practice, the binary samples are neither homogeneous nor large. Consequently, there are significant observational uncertainties in the measurement and comparison of these cutoff periods.

First, the observed cutoff periods are only lower limits to the true cutoff periods, as indicated in Figure 1. Due to sampling statistics it is unlikely that a binary with period equal to the cutoff period of the parent eccentricity distribution will be observed; however, given the definition of DMM, it is certain that a binary with a circular orbit will not have a period longer than the cutoff period. For a given theory of tidal circularization (and stellar interiors) and assumptions about the initial binary population (i.e., distributions of eccentricity, periods, primary and secondary mass, etc.), in principle a parent distribution of circularized orbits can be derived for any age. Given this distribution, the observed cutoff period could be taken as an estimator of the parent cutoff period rather than just as a lower limit. Indeed, with larger samples the distribution of eccentricity with period may be a more powerful discriminant than these simple diagnostics. A Monte Carlo study of the distribution of circularized orbits would be profitable for progress in this direction.

In any case, large samples of binaries are desired if the observed cutoff period is to provide a precise estimate of the parent cutoff period. After years of intensive observation, several of the samples - e.g., the Hyades, M67, field solar-mass and halo binaries - are approaching large sample sizes. Nonetheless, of these samples only M67 has a second binary with a circular orbit and period within 20% of the cutoff period given in Table 1. The cutoff periods are not well corroborated and thus are very sensitive to the specific nature of the binary having the longest period circular orbit. Similarly, the observed cutoff period for the Pleiades is derived from only two binaries with periods less than 15 days, both of which have circular orbits [DMM92]. These sampling issues demand caution in consideration of these data.

Note that, in principle, short-period eccentric orbits can also provide meaningful constraints on tidal circularization theories. For a given tidal circularization theory and binary age, the initial eccentricities of binaries with presently eccentric orbits can be computed [e.g., DMM92]. An implausible initial eccentricity distribution would cast doubt on a theory. For example, a very high tidal circularization rate might require an implausible number of binaries with very high initial eccentricities. Thus eccentric orbits may still play a role in setting upper limits on tidal circularization rates (and consequently on parent cutoff periods), but in a more complex, model-dependent way. In practice, the eccentric orbits in the existing binary samples do not meaningfully constrain the tidal circularization theories presently under consideration.

Second, direct comparison of observed cutoff periods presumes identical initial binary populations. (Note that the requirement is not that all binaries be the same but that the distributions of all relevant properties of a binary population be the same.) In practice, this is not the case for the binary samples in Table 1. An example is the halo sample. The stars comprising the halo binaries differ significantly from those in the cluster binary samples, particularly in chemical composition and primary mass. Thus arguably the halo cutoff period requires a correction based on both stellar interior and tidal circularization theory before comparison with the cluster results.

The cluster samples are also not homogeneous. As noted, a difficulty with the present definition of the cutoff period is that its value is set by one binary, and thus an observed cutoff period may be very sensitive to these inhomogeneities. For example, the Hyades cutoff period is set by the binary J331 [GGZ85]. This double-lined system consists of two $0.5\,M_\odot$ stars. As such, the primary is substantially different from the solar-mass primaries defining the cutoff periods in the Pleiades and M67, and perhaps should be excluded for the purposes of comparison with those samples. Whether the difference in mass and consequent differences in stellar structure (particularly in the nature of the convective zones) in fact significantly alters the tidal circularization timescale for J331 compared to solar-mass stars remains to be shown. In the absence of this analysis, we have chosen to include it for the definition of the Hyades cutoff. However, in Figure 1 we have also shown the Hyades cutoff period of 6.0 days for solar-mass stars alone.

These inhomogeneities are a fact of life; in practice binary samples are determined by magnitude limits, observing time and cluster richness as well as scientific goals. Hence detailed analyses of the binaries determining cutoff periods are merited for comparison of cutoff periods either between samples or with theory. At the least, the nature of the stars in the binary setting a cutoff period should be determined and noted by observers. With more effort, corrections for differences between samples can be calculated. For example, Mathieu and Mazeh [MaM88] explicitly convolved tidal circularization theory and stellar interior models of appropriate primary and secondary masses in order to interpret the observed cutoff periods of solar-mass stars in the Hyades and M67. Similar studies should be done for halo stars and J331, for example. Unfortunately, results of such calculations are necessarily dependent on the theories of stellar interiors, convection and tidal circularization. Hence, the more extreme the differences between samples, the less secure such corrections become.

4. Comparison of Observations and Theory

As briefly summarized in the Introduction, the theory of tidal circularization has been the subject of increased study in the last few years, in part motivated by the substantial increase in the number of binary orbit determinations. Nonetheless, all theories agree that the tidal circularization timescale T_c of a binary of period P can be expressed in the form $T_c = T_0(P/P_0)^\gamma$. The power-law dependence on period has two origins. If the dissipation mechanism is independent of orbital period, then dynamical considerations alone lead to $\gamma=16/3$ ([Zah77], [MaM88]). However, Zahn [Zah89] has argued that the dissipation efficiency also depends on period. In particular, he notes that the binaries with short enough periods to be circularized also have periods shorter than or comparable to the convective turnover timescale. Since convective viscosity is the source of dissipation, the consequence of these commensurate timescales is reduced dissipation and longer circularization timescales with shorter period compared to the classical theory. Zahn [Zah89] argued that this reduction in the turbulent viscosity depends linearly on period, so that $\gamma=13/3$, while Goldman and Mazeh [GoM91] found that the dependence is quadratic ($\gamma=10/3$). Finally, Tassoul ([Tas87], [Tas88]) finds $\gamma=49/12$ for his hydrodynamical mechanism for tidal circularization.

The term T_0 is a complicated function of the parameters describing the stellar structure and the nature of convection and turbulence. We will refer to it here as the "absolute rate calibration", a constant for a given type of star. It has recently come to the fore as Zahn and Bouchet [ZaB89], based on the theory of Zahn [Zah89], found the absolute rate calibration to be so small that main-sequence tidal circularization is insignificant. They conclude that essentially all circularization occurs during the early pre-main sequence phase of the binary components' evolution when the stars have much larger radii and deeper convective zones. Consequently, the circularization cutoff should be independent of binary age; Zahn and Bouchet compute this cutoff to be between 7.2 and 8.5 days for binaries with components between 0.5 M_O and 1.25 M_O. A consequence of this hypothesis is that γ cannot be determined from the distribution of cutoff periods with age. Indeed γ and T_0 could only be determined by comparison of the main-sequence cutoff period with calculations of orbit evolution through the pre-main sequence phase, and would be sensitively dependent upon presently poorly understood initial conditions and pre-main sequence evolutionary models.

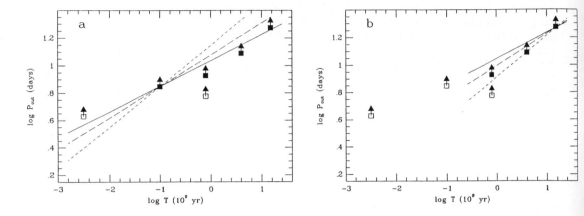

Figure 1. a) The distribution of cutoff periods with age (see Table 1). Also shown are the theoretical curves for $\gamma=16/3$ (solid; no reduction in dissipation with shorter period), $\gamma=13/3$ (long dashes; linear reduction with period) and $\gamma=10/3$ (short dashes; quadratic reduction with period). Open boxes are data discussed in the text but not used in evaluating the models (see Section 4.1). b) Same as a) except that Pleiades datum is not used in the analysis (Section 4.2).

For comparison of these theories with the observed distribution of cutoff periods, we first consider the classical hypothesis that tidal circularization throughout the main-sequence phase dictates the main-sequence cutoff periods. We will then reconsider the interpretation of the data in light of significant pre-main sequence tidal circularization.

4.1 Pure Main-Sequence Tidal Circularization

Since the presumption of effective tidal circularization throughout the main sequence phase is inconsistent with *a priori* calculations of T_0, we follow Goldman and Mazeh [GoM91] and use the data to calibrate the absolute rate calibration for solar-mass stars. In Figure 1a we show power laws with exponents 16/3, 13/3 and 10/3. (Note that the power law with exponent 49/12 is very close to that with exponent 13/3.) The observed cutoff period most displaced to larger values, after the time dependence of tidal circularization has been accounted for, is arguably the measured cutoff period nearest to the true cutoff period. As such, it represents the most significant lower limit on the rate of circularization. For these four main-sequence binary samples the Pleiades place such lower limits on the calibration of T_0 for all models. Consequently in Figure 1a we have normalized the models to the Pleiades cutoff period.[3] Of course, this method of normalization assures by construction that none of the models fall below the observed lower limits on the cutoff periods.

Given our normalization, we can investigate the dependence of tidal circularization on orbital period (i.e., the value of the exponent γ). The $\gamma=16/3$ model provides a good fit to the observed lower limits. If the true cutoff periods for all samples are not greatly in excess of their observed values, this model is supported by the data. On the other hand, the $\gamma=10/3$ model is most divergent from the observed lower limits. In particular, the $\gamma=10/3$ model predicts a true cutoff period in the Hyades of 13.1^d and in M67 of 21^d, uncomfortably high relative to the observed cutoff periods of 8.5^d and 12.4^d, respectively.

Thus, the observations are fully consistent with main-sequence tidal circularization being the primary influence setting the observed main-sequence cutoff periods. Furthermore, the observations do not demand reduction in the efficiency of convective viscosity and indeed, under the assumption of pure main-sequence tidal circularization, the observations favor little or no reduction in the rate of dissipation.

It should be recognized that this conclusion rests heavily upon the Pleiades cutoff period, which is derived from an uncomfortably small sample of binaries. As we shall see in the next section, with the (here arbitrary) exclusion of the Pleiades cutoff period the conclusions reverse and the data favor reduction in the efficiency of turbulent viscosity. Nonetheless, given the definition of cutoff period used here, the addition of more Pleiades orbits will not lower the cutoff period in future analyses.

Note that for the purpose of determining the period dependence of tidal circularization the pre-main sequence cutoff period shown in Figure 1a is not a significant constraint. As we shall discuss in the next section, if pre-main sequence orbital evolution produces a cutoff at the ZAMS which is greater than that predicted by extrapolating a main-sequence tidal circularization model to the ZAMS, then the cutoff period will simply not evolve until sufficient time has passed such that the shortest-period ZAMS eccentric orbits tidally circularize. Note though that if the ZAMS cutoff period is ≈ 4 days, as suggested by the limited pre-main sequence data in hand, then for the models shown in Figure 1a evolution of the cutoff period would in fact begin at or shortly after the ZAMS. Thus the presumption at the begining of this section that only main-sequence tidal circularization dictates main-sequence cutoff periods is consistent with present pre-main sequence observations.

4.2 Pre-Main Sequence and Main-Sequence Tidal Circularization

Zahn and Bouchet [ZaB89] have argued that significant tidal circularization occurs only during the pre-main sequence phase, primarily near the stellar birthline. Consequently main-sequence cutoff periods should be independent of age, with values of approximately 7-8 days according to their calculations. The specific cutoff period predicted by this hypothesis is necessarily dependent on both tidal circularization and pre-main sequence stellar evolution models, which are uncertain. Here we will first focus more generally on whether the observed distributions of eccentricities as a function of age are consistent with significant pre-main sequence tidal circularization.

---

[3] In both Figures 1a and 1b we have used the same normalization for all models for clarity in the figures. Precisely, in Figure 1a the $\gamma=16/3$ model should be normalized to the halo datum and in Figure 1b the $\gamma=10/3$ model should be normalized to the Hyades datum. In neither case do these small differences in normalization change the conclusions.

We consider first the hypothesis that pre-main sequence tidal circularization alone sets the observed main-sequence cutoff periods, so that the cutoff periods should be similar independent of age. In fact, the Pleiades and Hyades cutoff periods are similar despite an order of magnitude difference in age (and their values are indeed 7-8 days.) Nonetheless, older binary samples show longer cutoff periods. Circular orbits exist among field solar-mass binaries with periods as long as 10 days. The M67 binaries have a cutoff period of 12.4 days. The halo cutoff period is larger still at 18.7 days. These data indicate larger cutoff periods with increasing binary age, and as such are inconsistent with the hypothesis. Still, DMM have noted that pre-main sequence tidal circularization could produce circular orbits with periods as long as 10 days and not be inconsistent with the Pleiades and Hyades data, so that arguably the difference between the Pleiades, Hyades and M67 cutoff periods may not be as significant as it first appears.

Clearly a critical datum is the eccentricity distribution of zero-age main sequence or late pre-main sequence binaries. The pre-main sequence eccentricity distribution has been discussed by Mathieu [Mat92]; to date the longest period circular orbit has a period of 4.3 days. Thus there is no evidence that pre-main sequence circularization can establish cutoff periods as long as the Pleiades and Hyades, but similarly the data do not exclude that possibility. Two pre-main sequence binaries with eccentric orbits are of interest here. One has an orbital period of 7.5^d, an eccentricity of 0.3 and an age on the order of 10^6 yr. This binary does not exclude a cutoff period similar to that of the Pleiades and Hyades among pre-main sequence stars. However, it does indicate that circularization early in the pre-main sequence phase may not circularize binaries with periods longer than 7-8 days. Thus the increase in observed cutoff period in the older main-sequence binary samples is more likely significant. The second binary is the near-ZAMS star EK Cep at a period of 4.4 days with an eccentricity of 0.1. This binary suggests that pre-main sequence circularization does not extend as far the Pleiades and Hyades cutoffs, but there are significant caveats with this binary which weaken its significance [Mat92, DMM92].

It is also relevant to recall that the Hyades sample includes two binaries having substantial orbital eccentricities (e=0.35) at periods of only 5.8 days and 5.9 days. DMM have shown that these binaries can be naturally explained as incomplete circularization, but only if circularization rates are such that main-sequence tidal circularization is effective. The origin of these binaries is not clear given pre-main sequence circularization of periods at least as long as 7-8 days. Zahn and Bouchet make the *ad hoc* suggestion that their eccentric orbits are the product of post-main sequence evolution of the secondaries.

Thus, the increase in the cutoff periods of main-sequence binary samples with increasing age would argue against the hypothesis that pre-main sequence tidal circularization alone establishes all observed main-sequence cutoff periods. In addition, the existence of eccentric orbits at periods well below observed cutoff periods is not an immediate prediction of pre-main sequence tidal circularization, although the product of a range of eccentricities at the stellar birthline has not yet been considered. The most straightforward interpretation of both observations is that tidal circularization is effective on the main sequence, at least after ages of order 10^9 yr.

While the data argue for effective main-sequence tidal circularization, they do not exclude significant pre-main sequence tidal circularization as well. Torres *et al.* [TLM92] and DMM have suggested that both pre-main sequence and main-sequence tidal circularization play a role in creating the observed distribution of cutoff periods. For example, pre-main sequence evolution may indeed establish a cutoff period of 7-8 days, from which the

Pleiades and Hyades cutoff periods derive, and after the passage of order 10^9 yr main-sequence tidal circularization becomes effective at that orbital period. Indeed, the slope of the observed cutoff period distribution does steepen with increasing age, consistent with this idea. One advantage of this hybrid scenario is that the eccentric orbits of both EK Cep and the short-period Hyades binaries can still be attributed to incomplete tidal circularization, if main-sequence tidal circularization becomes significant at roughly the Hyades age [DMM92].

In this picture, the Pleiades cutoff period and possibly the Hyades cutoff period are attributed to pre-main sequence tidal circularization. Hence, the period dependence of the tidal circularization theory can only be determined from the oldest binary samples. In Figure 1b, we show the same three models presented in Figure 1a, but compared with only the three oldest binary samples. Clearly the cutoff periods of these older binaries delineate a steeper slope. Thus in this picture the data favor reduction in the dissipation efficiency of turbulent viscosity. Both the $\gamma=10/3$ and the $\gamma=13/3$ models provide close fits to the slope of the observed cutoff periods. Nonetheless, the fit of the $\gamma=16/3$ model is essentially the same as in Figure 1a and that model is not ruled out even given significant pre-main sequence tidal circularization.

Note that if the binary J331 were removed from the Hyades sample, the Hyades cutoff period would be lowered to 6 days (shown with an open box in Figure 1b; see Section 3). In this case, a quadratic dependence of the reduction on period and the consequent $\gamma=10/3$ period dependence of tidal circularization would be strongly supported by the data relative to the classical model. Additionally pre-main sequence tidal circularization would be required to establish a cutoff at somewhat lower periods ($\approx$6-7 days). Interestingly, this approach would also substantially reduce the difference in period between the shortest period eccentric orbit (5.7 days) and the cutoff period in the Hyades, so that none of the main-sequence samples would have eccentric orbits with periods much less than their cutoff periods. This would reduce the motivation for the significant period evolution during circularization on the main sequence suggested by DMM. Of course, the difference in cutoff periods among the older binary samples would be even more extreme, so that the primary argument for main-sequence tidal circularization would remain strong.

Thus to summarize, if pre-main sequence tidal circularization is sufficient to delay effective main-sequence tidal circularization until of order 10^9 yr, a reduction in the dissipation due to turbulent viscosity is supported by the data, as is the circularization theory of Tassoul. Unfortunately, the loss of dynamic range in age does not permit the data to clearly distinguish the extent of such reduction, or indeed select between the classical model of tidal circularization, models with reduced dissipation or the theory of Tassoul. The key issue in testing tidal circularization theories remains the extent of pre-main sequence tidal circularization, best addressed with further observation of young clusters and associations.

As a final point, it should be recognized that given the higher absolute rate calibrations of such hybrid models, the prediction by Zahn and Bouchet of a ZAMS cutoff period of 7-8 days is no longer relevant. For example, the greater absolute rate calibration required for significant main-sequence tidal circularization would likely lead to a larger ZAMS cutoff period (holding γ fixed at 13/3 as used by Zahn and Bouchet). Alternatively, other values of γ may shorten the predicted ZAMS cutoff period. More generally, any argument for effective main-sequence tidal circularization cannot ignore the consequent implications for tidal circularization during the pre-main sequence.

5. Conclusions

In principle, the distribution of circularization cutoff periods with age is a powerful means to constrain stellar tidal circularization theory. In practice, heterogeneity of binary samples, and in some cases insufficient samples sizes, somewhat weaken the observational constraint. Nonetheless, these cutoff periods represent an important interface between the observation and theory of tidal circularization.

Recently, theorists have introduced a significant modification to the classical theory of tidal circularization based on a period-dependent reduction in the efficiency of turbulent viscosity due to convective turnover timescales exceeding orbital timescales ([Zah89], [GoM91]). This reduction leads to smaller exponents in the power law dependence of circularization timescale with orbital period. Two new power laws have been suggested, depending upon differing theoretical arguments for the magnitude of the reduction. Alternatively, the fundamentally different hydrodynamical mechanism for tidal circularization of Tassoul [Tas87, Tas88] gives a power law dependence similar to those of the reduced dissipation theories.

We have found that the observed distribution of cutoff periods does not require this period-dependent reduction of circularization efficiency. The distribution can be well modeled by the classical hypothesis of effective tidal circularization throughout the main-sequence phase with little or no reduction. However, an interpretation of the data consistent with substantial reduction can be found if the cutoff periods in (only) the youngest main-sequence binary systems are presumed to be the result of pre-main sequence tidal circularization. In this picture, main-sequence tidal circularization is only significant in older clusters, for which models with reduced dissipation provide good fits to the observations. Similar conclusions can be drawn in regard to the theory of Tassoul.

The absolute calibration of tidal circularization rates have also been at issue, since recent theoretical derivations have found them to be such that main-sequence tidal circularization is insignificant. As a result, Zahn and Bouchet [ZaB89] concluded that the circular orbits found among main-sequence stars were established solely during the pre-main sequence phase of evolution, and thus circularization cutoffs should be independent of age. In fact main-sequence cutoff periods are observed to increase substantially with age, at least among clusters with ages greater than ≈ 1 Gyr. Also, in the Hyades and Praesepe clusters binaries are found with eccentric orbits at periods significantly less than the cutoff period. We suggest that the most straightforward interpretation of both results is effective tidal circularization during the main-sequence phase or, equivalently, a larger absolute rate calibration than found by theoretical arguments.

Nonetheless, significant pre-main sequence tidal circularization may still occur. The available observations cannot distinguish between pure main-sequence tidal circularization with the classical period dependence or a combination of pre-main sequence tidal circularization establishing cutoff periods ≈ 8 days in younger clusters with main-sequence tidal circularization producing further orbital evolution after the passage of ≈ 1 Gyr. On this ambiguity rests the inability of the distribution of observed cutoff periods to critically distinguish differing theories of tidal circularization. The essential observational task is better definition of the cutoff periods at ages near the zero-age main sequence in order to establish the extent of pre-main sequence tidal circularization. Eccentricity distributions for binary samples with ages less than 10^8 yr are critical.

Finally, in our analyses we have not addressed the significant uncertainties in comparison of cutoff periods due to inhomogeneous binary samples. Theoretical calculations to investigate the significance of the heterogeneities in the available binary samples would strengthen the interpretations of the extensive data already in hand.

RDM would like to gratefully acknowledge the support of National Science Foundation Grant AST8814986, the Presidential Young Investigator program and the Wisconsin Alumni Research Foundation. AD, MM and JCM acknowledge the constant support of the Fonds National Suisse de la Recherche Scientifique.

References

[DuM88] Duquennoy, A., Mayor, M. *Duplicity in the solar-neighborhood. III. New spectroscopic elements for nine solar-type binaries stars.* 1988, AA, 195, 129

[DuM91] Duquennoy, A., Mayor, M. *Multiplicity among solar-type stars in the solar neighbourhood. II. Distribution of the orbital elements in an unbiased sample.* 1991, AA, 248, 485

[DMM92] Duquennoy, A., Mayor, M., Mermilliod, J.-C. *Evolution of solar-mass binary orbital elements*, 1992, this volume (DMM)

[GoM91] Goldman, I., Mazeh, T. *On the orbital circularization of close binaries.* 1991, ApJ, 376, 260

[GGZ85] Griffin, R.F., Gunn, J.E., Zimmerman, B.A., Griffin, R.E. *Spectroscopic orbits for 16 more binaries in the Hyades field.* 1985, AJ, 90, 609

[JaM88] Jasniewicz, G., Mayor, M. *Radial velocity measurements of a sample of northern metal-deficient stars.* 1988, AA, 203, 329

[KoH81] Koch, R.H., Hrivnak, B.J. *On Zahn's theory of tidal friction for cool, main-sequence close binaries..* 1981, AJ, 86, 438

[LMC88] Latham, D.W., Mazeh, T., Carney, B.W., McCrosky, R.E., Stefanik, R.P., Davis, R.J. *A survey of proper-motion stars. VI. Orbits for 40 spectroscopic binaries.* 1988, AJ, 96, 567

[LMG92] Latham, D.W., Mathieu, R.D., Griffin, R.F., Milone, A.E., Davis, R.J., Mazeh, T. *Binaries in the old open cluster M67.* 1992, this volume

[LMT92] Latham, D.W., Mazeh, T., Torres, G., Carney, B.W., Stefanik, R.P., Davis, R.J. *The frequency, orbital characteristics and secondary masses of the main-sequence binaries in the halo and the disk.* 1992, this volume

[LWM76] Lecar, M., Wheeler, J.C., McKee, C.F. *Tidal circularization of the binary x-ray sources Hercules X-1 and Centaurus X-3.* 1976, ApJ, 205, 556

[Mat92] Mathieu, R.D. *The eccentricity distribution of pre-main sequence binaries.* 1992, this volume

[MLG90] Mathieu, R.D., Latham, D.W., Griffin, R.F. *Orbits of 22 spectroscopic binaries in the open cluster M67.* 1990, AJ, 100, 1859.

[MaM88] Mathieu, R.D., Mazeh, T. *The circularized binaries in open clusters: a new clock for age determination..* 1988, ApJ, 326, 256

[MWM89] Mathieu, R.D., Walter, F.M., Myers, P.C. *The discovery of six pre-main-sequence spectroscopic binaries.* 1989, AJ, 98, 987

[MaM84] Mayor, M., Mermilliod, J.-C. *Orbit circularization time in binary stellar systems.* 1984, in IAU Symposium 105, Observational Tests of Stellar Evolution Theory, eds. A. Maeder and A. Renzini (Dordrecht:Reidel), 411

[Maz90] Mazeh, T. *Eccentric orbits in samples of circularized binary systems: the fingerprint of a third star.* 1990, AJ, 99, 675

[MLM90] Mazeh, T., Latham, D.W., Mathieu, R.D., Carney, B.W. *On the orbital circularization of close binaries.* 1990, in NATO Advanced Study Institute, Active Close Binaries, ed. C. Ibanoglu, (Dordrecht:Kluwer), 145

[Sod83] Soderblom, D.R. *Rotational studies of late-type stars. II. Ages of solar-type stars and the rotational history of the Sun.* 1983, ApJS, 53, 1

[Tas87] Tassoul, J.-L. *On synchronization in early-type binaries..* 1987, ApJ, 322, 856

[Tas88] Tassoul, J.-L. *On orbital circularization in detached close binaries.* 1988, ApJ, 324, L71

[TLM92] Torres, G., Latham, D.W., Mazeh, T., Carney, B.W., Stefanik, R.P., Davis, R.J., Laird, J.B. *Tidal circularization among the close binaries in the halo.* 1992, in IAU Symposium 151, Evolutionary Processes in Interacting Binaries, eds. Y.Kondo, R.Sistero, R. Polidan, (Dordrecht:Reidel), in press

[Zah66] Zahn, J.-P. *Les marees dans une etoile double serree..* 1966, Ann. d'Ap., 29, 489

[Zah77] Zahn, J.-P.*Tidal friction in close binary stars..* 1977, AA, 57, 283

[Zah89] Zahn, J.-P. *Tidal evolution of close binary stars. I. Revisiting the theory of the equilibrium tide.* 1989, AA, 220, 112

[ZaB89] Zahn, J.-P., Bouchet, L. *Tidal evolution of close binary stars. II. Orbital circularization of late-type binaries.* 1989, AA, 223, 112

Concluding Remarks

Roger F. Griffin

It was only at the coffee break an hour ago that I was asked to say a few words now, and really a few words are all that I have wit enough to say at this juncture. The short notice has advantages for all of us: from my point of view I haven't had this great responsibility hanging over me throughout the Workshop, and from your standpoint you won't have to listen to a long harangue!

I will not try, as a sufficiently able final speaker often does (one who has been duly forewarned and therefore has listened with rapt attention to everything) to provide a systematic summary of what each of the previous speakers has said. Not only is it beyond my powers to do it, but it is also quite unnecessary – after all, you've all been to the Workshop, so you've heard everything for yourselves already! Instead, it is with great pleasure that I must say that the overwhelming impression that one has received here is of the splendid dynamism of our subject now. If it is thought worthy of celebration to produce a hundred orbits in as many papers over a long period of time, whatever are we to do when individual papers, such as seem imminently in prospect both from Geneva and from Harvard, carry a hundred or more new orbits?

I should think that this Workshop will go down in radial-velocity history (if at all) as the "$e - \log P$ conference". A remarkably large number of the speakers has come here armed, seemingly independently, with slides or viewgraphs showing for various groups of binaries exactly that relationship, which has not been a noticeably popular one in the past. I would be willing to make a small bet that the *Proceedings* of this conference will contain more $e - \log P$ plots than have ever been published previously in the whole history of the world! Of course it is one thing to show that a relationship exists, as in this case it clearly *does*, but it is quite another to understand what it actually means. I ask forgiveness from dissentients when I say that I do not believe that the last word has been said on that subject yet; but we need feel no great embarrassment on that score – it would be a dull world if problems were all so simple that they could be solved as soon as they were posed.

Being the last speaker offers the poetic justice of providing me with a ready-made platform from which to thank the organizers of this Workshop on behalf of all of us and perhaps particularly on my own behalf. We not only thank but congratulate Michel Mayor and Antoine Duquennoy on their superb and unusual choice of this venue and for their faultless organization. [Prolonged applause.] We are also much indebted to them, both for their perceptive choice of a topic that was clearly ripe for constructive discussion and also for not enforcing it so rigidly as to prevent certain speakers who shall be nameless from straying from it on occasion. On a more personal note, I should like to thank them very much for the inspired production, so unexpected, of the centenary cake, and for Dave Latham's excessively generous speech on that occasion. And on a still more personal note, please let me thank you very much, one and all, for coming to this conference and for being so kind to me both individually and collectively.